ADVANCES IN MOLECULAR AND CELLULAR ENDOCRINOLOGY

Volume 2 • 1998

ADVANCES IN MOLECULAR AND CELLULAR ENDOCRINOLOGY

Editor: DEREK LEROITH
Diabetes Branch
NIDDK
National Institutes of Health
Bethesda, Maryland

VOLUME 2 • 1998

 JAI PRESS INC.

Greenwich, Connecticut   *London, England*

CONTENTS

LIST OF CONTRIBUTORS

Veena R. Agarwal

The Cecil H. and Ida Green Center for
Reproductive Biology Sciences
University of Texas Southwestern
Medical Center
Dallas, Texas

Daniel G. Bichet

Department of Medicine
University of Montréal and Service
de Nephrologie
Montréal, Québec, Canada

Serdar E. Bulun

The Cecil H. and Ida Green Center for
Reproductive Biology Sciences
University of Texas Southwestern
Medical Center
Dallas, Texas

Wai-Yee Chan

Department of Pediatrics
Georgetown University, Children's
Medical Center
Washington, DC

Gordon B. Cutler, Jr.

National Institute of Child Health
and Human Development
National Institutes of Health
Bethesda, Maryland

James A. Fagin

Division of Endocrinology and Metabolism
University of Cincinnati College of Medicine
Cincinnati, Ohio

Fatima Ferrag INSERM Unit 344
Molecular Endocrinology
Paris, France

Vincent Goffin INSERM Unit 344
Molecular Endocrinology
Paris, France

Margaret M. Hinshelwood The Cecil H. and Ida Green Center for
Reproductive Biology Sciences
University of Texas Southwestern
Medical Center
Dallas, Texas

Paul A. Kelly INSERM Unit 344
Molecular Endocrinology
Paris, France

M. Dodson Michael The Cecil H. and Ida Green Center for
Reproductive Biology Sciences
University of Texas Southwestern
Medical Center
Dallas, Texas

Yuri E. Nikiforov Division of Endocrinology and Metabolism
University of Cincinnati College of Medicine
Cincinnati, Ohio

Alexander Oksche Rudolf-Buchheim-Institut für
Pharmakologie
Justus-Liebig-Universität Gießen
Gießen, Germany

Marc Reitman Diabetes Branch, NIDDK
National Institutes of Health
Bethesda, Maryland

Walter Rosenthal Rudolf-Buchheim Institut für
Pharmakologie
Justus-Liebig-Universität Gießen
Gießen, Germany

Peter Rotwein — Department of Molecular Medicine
Oregon Health Sciences University
Portland, Oregon

Alan R. Saltiel — Department of Cell Biology
Parke-Davis Pharmaceutical Research
Ann Arbor, Michigan

Evan R. Simpson — The Cecil H. and Ida Green Center for
Reproductive Biology Sciences
University of Texas Southwestern
Medical Center
Dallas, Texas

Michael J. Thomas — Department of Internal Medicine
University of Iowa College of Medicine
Iowa City, Iowa

Ying Zhao — The Cecil H. and Ida Green Center for
Reproductive Biology Sciences
University of Texas Southwestern
Medical Center
Dallas, Texas

PREFACE

Historically the field of endocrine research has always been at the forefront of scientific endeavors. The importance of these breakthroughs in research have been rewarded by numerous Nobel awards. In the field of diabetes alone, Nobel prizes have been awarded to researchers who discovered insulin, characterized the protein, and using insulin as a paradigm invented radioimmunoassays. Not surprisingly, therefor, biomedical researchers have always been attracted by the endocrine system and other similar systems of intercellular communication.

Over the past two decades, endocrine research has developed rapidly and adapted modern molecular and cellular biology techniques. These changes have allowed researchers in the filed to maintain their edge. Thus endocrine disease-related genes have been characterized and mutations in these genes have helped explain common and less common endocrine disorders. Our understanding of the regulation of gene expression has been greatly enhanced by molecular techniques.

In an attempt to bring investigators u to date with the recent advances in this exploding field we have decided to publish a yearly series on *Advances in Molecular and Cellular Endocrinology*. Internationally famous investigators have agreed to participate and their contributions hare appreciated. Each volume will include reviews on different aspects of endocrinology.

Volume 2 has focused on aspects of the pituitary gland both anterior (growth hormone and prolactin receptors, and GH action) and posterior (vasopressin) pituitary. In addition, thyroid cancer and steroidogenic enzymes and precocius puberty are covered. Finally, the "hot topics" include leptin and growth factor signaling. Once again I have tried to include a diverse array of topic and hope that readers benefit as much as I have when reading this second volume.

Derek LeRoith
Editor

Chapter 1

Molecular Aspects of Prolactin and Growth Hormone Receptors

VINCENT GOFFIN, FATIMA FERRAG, and

PAUL A. KELLY

Advances in Molecular and Cellular Endocrinology
Volume 2, pages 1-33.
Copyright © 1998 by JAI Press Inc.
All rights of reproduction in any form reserved.
ISBN: 0-7623-0292-5

INTRODUCTION

Prolactin (PRL) and growth hormone (GH) receptor (R) cDNAs were cloned 10 years ago (Leung et al., 1987; Boutin et al., 1988). Both are single-pass transmembrane chains that display a relatively low degree (~ 30%) of similarity (Kelly et al., 1991). Initially, these receptors were believed to form a new family since their DNA and amino acid sequences failed to exhibit any homology with any other known receptor. However, further sequence comparison with newly-identified membrane receptors led to the identification of a new family of receptors sharing several structural and functional features with PRLR and GHR (Bazan, 1989, 1990; Kelly et al., 1991). Termed Class-1 cytokine receptors, this new superfamily includes receptors for several interleukins, granulocyte-colony stimulating factor (G-CSF), granulocyte macrophage-colony stimulating factor (GM-CSF), leukemia inhibitory factor (LIF), oncostatin M (OM), erythropoietin (EPO), thrombopoietin (TPO), gp130 and the obesity factor, leptin (Stahl and Yancopoulos, 1993; Taga and Kishimoto, 1993; Kitamura et al., 1994; Finidori and Kelly, 1995; Goffin et al., 1996b). The newly established link between all these receptors has greatly facilitated functional studies at the molecular level, since it rapidly became clear that cytokine receptors share several features in the mechanisms of activation and of signal transduction. This review is aimed at summarizing the current knowledge on the molecular aspects of PRLR and GHR. The first part focuses on the extracellular, ligand binding domain, whereas the second part is centered on the intracellular signalling domain. We conclude this article with a discussion of physiological significance of the different receptor isoforms.

THE EXTRACELLULAR DOMAIN

Structure

In the rat, three different isoforms of the PRLR have been isolated (Boutin et al., 1988; Shirota et al., 1990; Ali et al., 1991). Although they differ in length, composition and biological properties (see Figure 1 and The Intracellular Domain), their extracellular domains are identical and encompass the 210 amino-terminal residues. The primary structure of the GHR extracellular domain is similar and is composed of 246 amino acids (aa) in humans (Leung et al., 1987). Most of the sequence similarities between cytokine receptors lie within the extracellular domain, which is typically composed by one domain of ~ 200 aa that can be divided into two subdomains of ~ 100 aa (referred to as D1 and D2; see Figures 1 and 2A), showing each analogy with the type III module of fibronectin (Bazan, 1990; Kelly et al., 1991). Although some cytokine receptors contain additional domains, it seems that ligand interactions are primarily driven by the conserved fibronectin-like domains (Kossiakoff

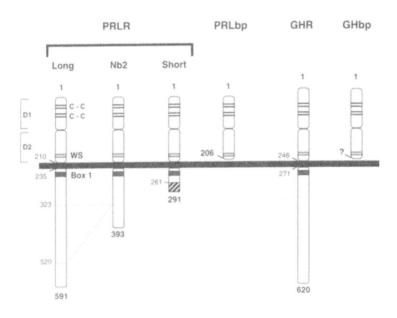

Figure 1. PRL and GH receptors. The three isoforms of rat PRLR are represented (Boutin et al., 1988; Shirota et al., 1990; Ali et al., 1991). The Nb2 PRLR differs from the long isoform by a 198 aa deletion in the cytoplasmic domain. Otherwise, the short PRLR is identical to both other isoforms up to residue 261, after which its sequence differs (hatched box). In the human breast cancer cell line BT-474, an mRNA encoding a soluble PRLBP of 206 aa has been isolated (Fuh and Wells, 1995). Soluble hGHR is produced by proteolysis of membrane hGHR (Leung et al., 1987) at a site that remains unidentified. In the extracellular domains, two pairs of disulfide bonded-cysteines and the WS motif (Trp-Ser-X-Trp-Ser) are characteristic features of the cytokine receptor superfamily, although conservative substitutions are found in the GHR (see text). Subdomains D1 and D2 of the extracellular domain are indicated.

A

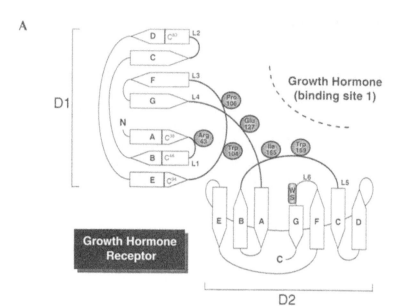

B

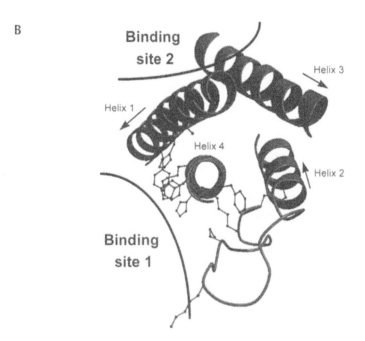

continued

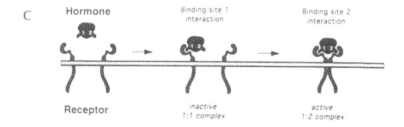

Figure 2. Hormone-receptor interaction : structure and binding sites. (**A**) Receptor extracellular domain. The secondary structure of the extracellular domain of PRL and GH receptors (binding proteins) is schematically represented. Binding proteins contain two subdomains (D1 and D2), each of which is composed of seven β-strands (symbolized by open boxes; arrowheads indicate the N- to C-terminal orientation) that fold in two antiparallel β-sheets containing strands A, B, and E for the first, and strands C, D, F and G for the second. Loops 1 to 6 (L1 to L6) link pairs of β-strands (De Vos et al., 1992; Somers et al., 1994). Crucial residues for ligand binding (some of which are indicated) are mainly located on loops, especially loops L1, L3 and L5. In the PRLR, Trp 72 and 139 are homologous to GHR Trp 104 and 169, respectively. (**B**) Hormone 3D structure. Three dimensional structure of hPRL modeled using the crystallographic coordinates of porcine GH (Abdel-Meguid et al., 1987; Goffin et al., 1995). These hormones fold into a four-helix bundle. The location of binding sites 1 and 2 is indicated and residues involved in PRL binding to the PRLR at the binding site 1 interface that have been determined by site-directed mutagenesis are represented (Goffin et al., 1992; Goffin et al., 1996b; Kinet et al., 1996). (**C**) Ligand-induced receptor homodimerization. Hormone binding to PRLR and GHR is sequential. First, the hormone interacts with one molecule of receptor through its binding site 1, forming an inactive 1:1 complex. Then, the hormone binds to a second (and identical) molecule of receptor through its site 2 to reach an active 1:2 complex (receptor homodimerization). Engineered hormone analogues whose binding site 2 is sterically blocked are unable to induce receptor homodimerization and are inactive. Due to their ability to bind to through site 1, such analogues block the receptors in the inactive 1:1 conformation and act as antagonists of wild type hormones.

et al., 1994). Two highly conserved features are found in the extracellular domain of cytokine receptors. The first one is two pairs of disulfide-linked cysteines in the N-terminal subdomain (Figure 1; see Two Pairs of Cysteines, below). Although the PRLR and GHR show only limited overall sequence identity, the aa bordering these disulfide-bonded cysteines display up to 70% identity, suggesting a functional involvement of these regions (see below). The second typical feature is a pentapeptide, termed "WS motif" because of its amino acid composition (Trp-Ser-any amino acid-Trp-Ser, or WSxWS in one letter code; Figure 1 and see The WS Motif, below) and found in the membrane proximal region of the C-terminal subdomain (D2) of the extracellular domain. Remarkably, the GHR is the only class-1 cytokine receptor for which the WS motif is not exactly replicated, three of the four consensus aa being conservatively mutated (YGeFS substituted for WSxWS).

The three-dimensional (3D) structures of genetically engineered human (h) GHR and hPRLR extracellular domains have been determined by crystallographic analysis (De Vos et al., 1992; Kossiakoff et al., 1994; Somers et al., 1994). Despite their low degree of identity (28%), they exhibit the same tertiary structure. In both receptors, each fibronectin-like subdomain (D1 and D2; Figure 2A) contains seven β-strands that fold in a sandwich formed by two anti-parallel β-sheets, one composed by 3 strands (referred to as strands A, B, and E according to Somers and colleagues (1994)) and the other by four strands (C, D, F, and G) (De Vos et al., 1992; Kossiakoff et al., 1994; Somers et al., 1994). Both subdomains are linked by a small four residue polypeptide (De Vos et al., 1992; Somers et al., 1994), and six loops (L1 to L6; Figure 2A) linking pairs of β-strands have been defined (De Vos et al., 1992; Somers et al., 1994). With the exception of some differences in the relative orientations of the N- and C-terminal domains with respect to each other, the 3D structures of PRLR and GHR extracellular domains are virtually identical (Kossiakoff et al., 1994). As anticipated (Bazan, 1990), this folding pattern is likely to be shared by several, if not all, cytokine receptors, since it has also been described for the EPOR extracellular domain (Caravella et al., 1996; Livnah et al., 1996) and for the α chain of the interferon-γ receptor, a class-2 cytokine receptor (Walter et al., 1995).

Ligand Binding

In addition to the membrane-anchored receptors, soluble PRL- and GH-binding proteins (BP) have been identified (see Physiological Role of the Receptor Isoforms) (Leung et al., 1987; Postel-Vinay et al., 1991; Fuh and Wells, 1995). These GHBP and PRLBP are produced either by alternative splicing of their respective full-length receptor primary transcript, or by proteolytic cleavage of membrane-bound receptors (Baumbach et al., 1989; Sotiropoulos et al., 1993); in any case, binding proteins are identical to the extracellular domain of membrane receptor whose structure has been determined (see above). This demonstrates that the extracellular domain of these receptors is necessary and sufficient for binding to their natural ligand, PRL and GH.

The first crystallographic study of ligand-binding protein complex in the PRL/GH family led to the remarkable discovery that one molecule of hGH is bound to two molecules of hGHBP (De Vos et al., 1992). The interaction involves two distinct regions of the ligand, referred to as binding sites 1 and 2, each of which interacts with a single BP (Cunningham et al., 1991; De Vos et al., 1992; Goffin et al., 1996b; Wells, 1996) (Figure 2B). Interestingly, although the binding sites on the hormone are asymmetrical, they interact with similar regions on the receptor, even though contact residues are not strictly identical (De Vos et al., 1992; Kossiakoff et al., 1994). To date, crystallogenesis of PRL-PRLBP complexes have been unsuccessful. However, primate GHs exhibit the unique ability among GHs to bind to the PRLR and crystals of hGH-hPRLBP complexes have been recently reported (Somers et al., 1994). Contrary to what is observed for the hGH-hGHBP complexes, crystals of hGH-hPRLBP are composed of only one molecule of each kind (the hGH binding site 2 does not interact with a second hPRLBP molecule in the crystal). The 3D structures of the hGH-GHBP and hGH-PRLBP 1:1 complexes appear almost superimposable and only minor differences are observed in terms of binding epitopes and relative orientation of proteins, although the two types of interaction involve different subsets of aa on both the ligand and the receptor (De Vos et al., 1992; Kossiakoff et al., 1994; Somers et al., 1994; Goffin et al., 1996b). PRL is predicted to adopt a GH-like four-helix bundle folding (Abdel-Meguid et al., 1987; Goffin et al., 1995). The structural analogies between, on the one hand, PRL and GH and, on the other hand, their respective receptors, strongly suggest that receptor-ligand interactions within the PRL/GH family occur through a very similar mechanism (Goffin et al., 1996b).

Typically, such protein-protein interactions involve a large set of aa Although crystallographic studies have permitted the identification of contact residues (De Vos et al., 1992; Somers et al., 1994), mutational analysis has been a powerful approach to evaluate the individual involvement of any amino acid in the binding or biological properties of both the receptors and their ligands. Three typical features of PRLR and GHR have been particularly studied using this approach: the disulfide-linked cysteine loops, the WS motif and two tryptophane residues located in between these two conserved features. In the following sections, we attempt to correlate data obtained by site-directed mutagenesis with more recent investigations of the structural and energetical aspects of hormone-receptor complexes.

Two Pairs of Cysteines

In both hGHBP and hPRLBP, β-strands A and B of the N-terminal domain (D1) are bridged by a disulfide link (Cys 38–48 for GHR, Cys 12–22 for PRLR), as are β-strands D and E of the same domain (Cys 83–94 for GHR, Cys 51–62 for PRLR) (De Vos et al., 1992; Somers et al., 1994) (Figure 2A). As frequently assumed for disulfide bridges, the major role of these features is likely to ensure a proper folding of the extracellular domain. Accordingly, mutation of these cysteines in the PRLR leads to impaired structural and functional properties of the receptor (Rozakis-Adcock and Kelly, 1991). Each disulfide loop borders regions of higher similarity compared to the rest of the extracellular domain of PRLR/GHR (Kelly et al., 1991; Rozakis-Adcock and Kelly, 1992), which suggests a functional involvement of some of the aa in between these cysteines. In agreement, site-directed mutagenesis revealed the implication in ligand binding of residues within the first, but not the second disulfide-loop for both receptors (Bass et al., 1991; Rozakis-Adcock and Kelly, 1992). This first cysteine loop corresponds to loop L1 and carries some residues (e.g., Arg 43, Glu 44) whose implication in ligand binding has been confirmed by structural analysis (De Vos et al., 1992; Somers et al., 1994; Clarckson and Wells, 1995) (Figure 2A).

Trp 104 and 169

With respect to the GHR, a site-directed mutagenesis study has identified one Trp residue (Trp 104) as crucial for efficient ligand binding since its individual mutation drops the affinity for hGH by a factor of 2500 (Bass et al., 1991). Other studies also pointed out the key role of another aromatic residue, Trp 169 (De Vos et al., 1992; Clarckson and Wells, 1995). Trp 104 and 169 are located on loops L3 and L5, respectively (Somers et al., 1994) (Figure 2A). In the homodimerized hGH-hGHBP complex (De Vos et al., 1992), the tryptophans of both receptor molecules are involved in interactions with both binding site interfaces of the ligand and are buried in hydrophobic environments formed by residues identified as hGH binding determinants (Cunningham and Wells, 1989, 1991; Cunningham et al., 1991; De Vos et al., 1992; Somers et al., 1994).

Interestingly, these two Trp residues are conserved in the PRLR (Trp 72 and Trp 139) and comparison of hydrogen bonds and salt bridges at the binding site 1 interface of hGH-hGHBP and hGH-hPRLBP complexes (De Vos et al., 1992; Somers et al., 1994) revealed that the hydrogen bond involving the first Trp (104 or 72, respectively) and hGH Lys 168 is the sole conserved in both crystal structures. This interaction is also anticipated to be maintained in the hPRL-hPRLBP complex since hGH Lys 168 is replaced by an Arg 177 in hPRL (Goffin et al., 1995), although this assumption awaits for structural determination of the complex.

Remarkably, despite the low sequence similarity of extracellular domains of PRL and GH receptors, the distribution along the molecule of the residue patches involved in hGH binding is almost identical (De Vos et al., 1992; Somers et al., 1994) and mainly involves the six loops (L1 to L6) linking β-strands (Figure 2A). Energetic study of hGH-hGHBP interface identified a functional epitope (Clarckson and Wells, 1995), defined by a hot spot of binding energy and involving a much restrained number of aa than the structural epitope, defined by the contact residues with the ligand (De Vos et al., 1992). This functional epitope mainly encompasses loops L1 (residues Arg 43 and Glu 44), L3 (residues Trp 104 and Pro 106), and L5 (residues Ile 165 and Trp 169) (Figure 2A) and appears highly conserved in the hGH-hPRLBP interaction (Somers et al., 1994). More remarkably, the involvement of these loops in ligand binding is also conserved in a peptide-EPOR complex (Livnah et al., 1996), despite low sequence homology between either ligands or receptors. This strengthens the assumption that hormone-receptor interactions within the class-1 cytokine family are driven by very similar rules, even if interactions involving some key residues such as Trp 104 and 169 are likely features typical of the PRL/GH receptor family.

The WS Motif

As described above, the WS motif is found per se in all cytokine receptors with the exception of the GHR. Although such remarkable conservation within a family of receptors otherwise displaying very low degree of similarity argues for a major functional implication of this feature, the exact role of the WS motif remains an enigma. In crystal structures of ligand-bound GHR (De Vos et al., 1992), PRLR (Somers et al., 1994), or EPOR (Caravella et al., 1996; Livnah et al., 1996), the WS motif is located away from the binding interfaces and is thus unlikely to be involved in any ligand interaction. Curiously, however, mutation within the WS motif of PRLR (Rozakis-Adcock and Kelly, 1992), GHR (Baumgartner et al., 1994), EPOR (Quelle et al., 1992), or IL-2Rβ (Miyazaki et al., 1991) leads to reduction of ligand binding affinity. One explanation might be that this feature participates in maintaining the overall structure of the extracellular domain (Livnah et al., 1996). Accordingly, several authors reported some modifications in the cellular trafficking of WS mutants (passage from endoplasmic reticulum to Golgi apparatus, cell surface expression, internalization, etc.), which usually correlates with structural disturbance (Miyazaki et al., 1991; Quelle et al., 1992; Baumgartner et al., 1994; Hilton et al., 1996). Alternatively, the possible involvement of the WS motif in an interaction with an accessory protein has also been proposed (Miyazaki et al., 1991; Rozakis-Adcock and Kelly, 1992), but awaits experimental demonstration.

Natural mutation of the WSxWS motif into the YGeFS sequence in the GHR remains intriguing. Crystal structures show that the overall conformation of the WS motif is identical between PRLR and GHR (Somers et al., 1994; Livnah et al., 1996). Engineered substitution of the consensus WS motif for the modified WS in GHR does not affect the behavior of the receptor in term of binding, cellular localization or biological properties (Baumgartner et al., 1994). This suggests that the aa of the modified WS motif within the GHR neither result from adaptive mutations nor are required per se for GHR function. Otherwise, alanine substitutions of the last serine and, to a lower extent, of the first tyrosine in the GHR WS motif (YGeFS) are detrimental to binding, bioactivity and cell membrane localization, presumably due to tertiary structure alteration (Duriez et al., 1993; Baumgartner et al., 1994). This correlates with the identification of a naturally-occurring mutation of the most C-terminal serine into an isoleucine in the GHR WS motif of a chicken strain exhibiting a sex-linked dwarf phenotype (Duriez et al., 1993). Although it remains premature to draw general conclusions for all cytokine receptors, it is reasonable to postulate from available data that the consensus WS motif of PRLR/GHR requires an aromatic in the first position and a serine in last position.

Receptor Homodimerization Leads to Receptor Activation

The observation that hGH-hGHBP crystals are formed of complexes containing one hGH molecule bound to two hGHBP molecules strongly suggested the occurrence of ligand-induced dimerization of the membrane-anchored receptor. In agreement, functional studies of GHR showed first, that membrane GHR is activated by GH-induced dimerization and second, that receptor homodimerization occurs sequentially, the interaction via hormone binding site 1 being required before interaction can occur via binding site 2 (Fuh et al., 1992, 1993) (Figure 2B and C). We (Goffin et al., 1994, 1995, 1996a) and others (Fuh et al., 1993; Hooper et al., 1993; Fuh and Wells, 1995) have provided several lines of evidence to suggest that ligand-induced activation of the PRLR also occurs through dimerization and follows the sequential model initially described for hGH (Fuh et al., 1992), although no crystal structure of PRL-PRLBP complexes is available yet to confirm this assumption. This is, however, in good agreement with the general rule that cytokine receptors are activated by oligomerization of two or more membrane polypeptide chains that are either identical or represent different subunits (Stahl and Yancopoulos, 1993; Taga and Kishimoto, 1993; Kitamura et al., 1994; Finidori and Kelly, 1995; Goffin et al., 1996b).

Analysis of the hGH-hGHBP crystal also revealed the occurrence of contacts between each binding protein at the receptor interface (De Vos et al., 1992). In the context of membrane-anchored receptors, the functional necessity of contacts at the receptor homodimer interface (along the extracellular, transmembrane and/or cytoplasmic domains) remains unknown. It should be noted, however, that mutations in the extracellular domain at the receptor-receptor interface have been documented in patients presenting clinical features of Laron syndrome, a disease that is characterized by GH resistance and results in dwarfism (Duquesnoy et al., 1994). This suggests that more than only hormone-receptor interactions are required to achieve a fully active transducing complex. As another illustration of the importance of the receptor-receptor contacts, it has been reported recently that deletion of the second fibronectin-like domain (D2) in the PRLR leads to a constitutively active receptor, presumably due to spontaneous receptor dimerization (Gourdou et al., 1996). Finally, it has been suggested that these receptor interface contacts are typical of class-1 cytokine receptors since they are not observed in the α chain homodimer of the class-2 IFN-γ receptor (Walter et al., 1995).

Based on the model of receptor activation by ligand-induced homodimerization, PRL and GH antagonists have been designed (Fuh et al., 1992, 1993; Fuh and Wells, 1995; Goffin et al., 1996a,b). As described above, the functional binding epitope involves two conserved Trp residues on receptors (Figure 2A). At binding site 2, these Trp bury into a cleft delimited by helices 1 and 3 on the hormone (Wells, 1994; Goffin et al., 1996b) (Figure 2B). Several recent studies have demonstrated that a single mutation introducing steric hindrance within the hormone binding site 2 (e.g., glycine replaced by arginine) abolishes interaction with a second receptor molecule, presumably by preventing proper docking of the Trp residues within the cleft (Fuh et al., 1992, 1993; Fuh and Wells, 1995; Goffin et al., 1996a). As a result, such hormone analogues fail to induce receptor homodimerization and are thus inactive. However, since they maintain the ability to bind to the receptor through their binding site 1, they block the receptor in the inactive 1:1 stoichiometry and thus act as hormone antagonists. In other words, development of antagonists has resulted from the separation of two naturally successive events, i.e., hormone-receptor binding and receptor homodimerization.

Although the development of PRL/GH antagonists is just beginning, it is promising for future therapeutic applications. For example, excess of GH secretion leads to several metabolic disorders, gigantism and acromegaly among them. Administration of GH antagonists able to block the GHR in an inactive stoichiometry and competing for endogenous GH binding, could lower, or even prevent, the occurrence of such pathologies. To our knowledge, no pathology has been clearly linked to PRL disorders. However, Vonderhaar and colleagues have recently reported that the human breast cancer cell line T-47D secretes high amounts of hPRL which, in turn, exerts an autocrine/paracrine effect on cell proliferation (Ginsburg and Vonderhaar, 1995). If such events actually occur in breast tumor cells *in vivo*, availability of PRL antagonists (Fuh and Wells, 1995; Goffin et al., 1996a) might be an alternative clinical approach in treatment of some forms of breast cancer.

THE INTRACELLULAR DOMAIN

Structure

The hGHR cDNA encodes a mature protein composed of 620 aa divided into an extracellular (246 aa), a transmembrane (24 aa), and a long cytoplasmic (350 aa) domain (Leung et al., 1987) (Figure 1). Three PRLR isoforms have been identified from rat tissues and they only differ in their cytoplasmic domains (Kelly et al., 1991). The first form of PRLR was isolated from a liver cDNA library (Boutin et al., 1988) and encodes a protein with a short cytoplasmic domain (57 aa). A second PRLR isoform with a longer cytoplasmic tail (357 aa), and therefore referred to as "long" form by comparison to the "short" form initially identified, was further isolated from rat ovary and liver (Shirota et al., 1990). These two forms are encoded by the same gene and result from alternative splicing (Shirota et al., 1990); they are strictly identical until residue 261, after which their sequences diverge. Finally, a mutant of the long PRLR isoform has also been identified in the rat Nb2 lymphoma cell line (Ali et al., 1991). Intermediate in size (393 aa), the Nb2 PRLR results from a 594 base pair deletion in the region of the gene encoding the cytoplasmic domain (Ali et al., 1991). With the exception of the resulting 198 aa deletion, the Nb2 PRLR is thus identical to the long PRLR isoform (Figure 1).

To the best of our knowledge, no structural data has yet been reported for the cytoplasmic domain of any cytokine receptor. The cytoplasmic domains of PRLR and GHR, and of cytokine receptors in general, display only limited sequence similarity. Two regions, called Box 1 and Box 2 (Kelly et al., 1991; Murakami et al., 1991), are relatively conserved. Box 1 is a membrane-proximal region composed of 8 aa highly enriched in prolines and hydrophobic residues. Due to the particular structural properties of proline residues, the conserved PxxP (proline - any aa - any aa - proline) motif within cytokine receptor Box 1 is assumed to adopt a consensus folding specifically recognized by some transducing molecules (see below). Although proline-rich regions are frequently regarded as Src homology (SH)3 domain binding entities (Ren et al., 1993), recent structural analysis of the PRLR proline rich motif has put this hypothesis into question (O'Neal et al., 1996). However, functional study of Box 1 unambiguously identifies this conserved region as an absolute requirement for cytokine receptor signaling. The second consensus region, Box 2, is much less conserved than Box 1 and consists in the succession of hydrophobic, negatively-charged then positively-charged residues.

Signal Transduction Pathways

Receptor-Associated Kinases

JAK2. It was observed several years ago that hormonal stimulation of PRLR or GHR leads to tyrosine phosphorylation of several cellular proteins, including the receptors themselves (Carter-Su et al., 1989; Rui et al., 1992). Since all cytokine receptor cytoplasmic domains are devoid of any intrinsic enzymatic activity, such phosphorylation had to result from one (or several) associated protein kinase(s) also activated upon hormonal stimulation. The first major step in the understanding of PRLR/GHR signaling (Figure 3) was the identification of JAK2 as one of these tyrosine kinases (Argetsinger et al., 1993; Campbell et al., 1994; Lebrun et al., 1994; Rui et al., 1994). JAK2 belongs to the Janus tyrosine kinase family, a family which currently includes four members (JAK1,

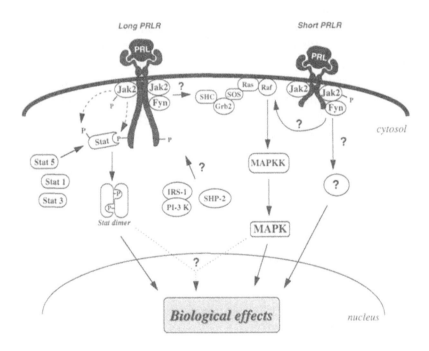

Figure 3. Schematic representation of some PRL and GH receptor signaling pathways. Long and short PRLR isoforms are represented. PRLR and GHR activate STAT1, 3, and 5. The differential involvement of STAT proteins in pathways leading to activities related to cell differentiation or proliferation remains poorly understood. Whether the short PRLR isoform activates the STAT pathway is currently unknown. The MAPK pathways is activated by both GH and PRL receptors (probably including short PRLR isoform). Interactions between receptors and Src kinases, SHP-2 phosphatase and other transducing molecules remains unclear.

JAK2, JAK3, and Tyk2) that are all involved in signaling of cytokine receptors (Taga and Kishimoto, 1993; Kishimoto et al., 1994; Ihle, 1994; Finidori and Kelly, 1995; Ihle and Kerr, 1995). With respect to PRLR and GHR, JAK2 is the major receptor-associated Janus kinase, although involvement of JAK1 has also been proposed (Dusanter-Fourt et al., 1994; Smit et al., 1996).

JAK2 is constitutively associated with the PRLR (Lebrun et al., 1994), whereas the GHR associates with the kinase only upon ligand activation (Argetsinger et al., 1993). This difference suggests that some ligand-induced conformational changes of the GHR might be required prior to JAK2 binding. For both receptors, it has been demonstrated that the membrane-proximal Box 1 is absolutely required for the receptor-JAK2 interaction (Sotiropoulos et al., 1994; Lebrun et al., 1995b; Dinerstein et al., 1995), although requirement of additional residues towards the C-terminus cannot be ruled out (DaSilva et al., 1994, 1996). As mentioned above, Box 1 is highly enriched in proline residues and is believed to adopt the typical folding of SH3-binding domains (Ren et al., 1993). Janus kinases are devoid of SH3 motifs, suggesting that the interaction, if it is direct, involves a mechanism different from the known SH3-SH3 binding domain. A recent mutational study of JAK2 identified the N-terminal fifth of the kinase as the region involved in GHR binding (Frank et al., 1995); whether this region binds directly to Box 1 remains unknown. Alternatively, the interaction might be mediated by an adapter protein containing an SH3 domain. A recent biochemical study of synthetic peptides corresponding to the Box 1 sequence of PRLR has brought a new insight on these hypotheses (O'Neal et al., 1996). Using spectrometric approaches (circular dichroism, nuclear magnetic resonance), O'Neal and colleagues proposed that a Box 1 peptide adopt a folding different from reported SH3 binding domain and, therefore, can not interact with SH3 domain protein. Instead, they suggest that isomerization of the second proline of the consensus PxxP motif regulates the activity of the receptor in an "on/off" switch manner (O'Neal et al., 1996). Although assessment of these hypotheses requires further experimental investigations, the proposed functional importance of the second proline residue of the PxxP motif correlates with mutational data (Dinerstein et al., 1995; Wang and Wood, 1995).

The mechanism by which JAK2 is activated remains poorly understood. It is usually assumed that ligand-induced oligomerization of cytokine receptors brings two JAK molecules close to each other, which allows trans-phosphorylation on tyrosines and subsequent activation of the enzyme (Finidori and Kelly, 1995; Ihle and Kerr, 1995). In the PRLR/GHR context, activation of JAK2 occurs very rapidly after hormonal stimulation (within 1 minute), strongly suggesting that this event is upstream of several signalling pathways of both receptors. Accordingly, receptor mutants in which Box 1 has been deleted, or in which proline residues have been mutated, can neither associate with nor activate JAK2 and were reported inactive in all *in vitro* biological responses analyzed so far (Colosi et al., 1993; DaSilva et al., 1994; Goujon et al., 1994; Dinerstein et al., 1995; Lebrun et al., 1995b; Sotiropoulos et al., 1995; Wang and Wood, 1995). The sole exception is the GH-dependent increase in intracellular Ca^{2+} (Billestrup et al., 1995) that seems independent of JAK2 activation.

Src Kinases.   Src kinases form another family of tyrosine kinases. Contrary to Janus kinases, they contain one SH2 and one SH3 domain and are thus candidates for binding to phosphotyrosine or proline-rich regions (Erpel and Courtneidge, 1995). Fyn, a member of this kinase family, is associated with the PRLR and activated by PRL stimulation in the rat T lymphoma Nb2 cell line (Clevenger and Medaglia, 1994). Association of the PRLR with Src, the prototype member of this kinase family, has also been reported after PRL stimulation in lactating rat hepatocytes (Berlanga et al., 1995). Recently, insulin receptor substrate-1 (IRS-1) has been shown to be associated with and phosphorylated by Fyn (Sun et al., 1996), itself a substrate of the GHR (Argetsinger et al., 1995); this suggests possible involvement of the Fyn kinase in GHR signaling. The role of Src kinases in signal transduction by PRL/GH receptors remains unknown, although promotion of cell growth has been suggested by some authors (Berlanga et al., 1995).

Receptor Phosphorylation

When activated, JAK2 phosphorylates the receptor on tyrosines (Sotiropoulos et al., 1994; Lebrun et al., 1995a; VanderKuur et al., 1995b; Wang et al., 1995). Phosphotyrosine are binding sites for SH2 domains and, therefore, play a central role in the interactions between transducing proteins. Efforts have thus been made to map the cytoplasmic tyrosines of PRLR and GHR undergoing phosphorylation upon hormonal stimulation.

In the intermediate (Nb2) PRLR isoform, which contains only 3 tyrosine residues, we have identified the most C-terminal (Tyr 382) as the sole tyrosine undergoing phosphorylation upon receptor/JAK2 activation (Lebrun et al., 1995a). When a Phe is substituted for the equivalent C-terminal tyrosine in the long PRLR isoform, which contains six additional tyrosines, receptor phosphorylation still occurs. This indicates that at least one of the eight remaining tyrosine residues is also a substrate of JAK2. Interestingly, although the short rat PRLR contains four tyrosines and activates JAK2 (Lebrun et al., 1995b), this isoform does not undergo phosphorylation upon stimulation.

The GHR also contains several tyrosine residues (nine for the rabbit GHR). Using different truncated GHR, we have shown that the five C-terminal tyrosines are major phosphorylation sites (Sotiropoulos et al., 1994, 1996). This remains controversial, however, since other investigators also mapped the membrane-proximal Tyr 333 and Tyr 338 as being phosphorylated (Hackett et al., 1995; Lobie et al., 1995; VanderKuur et al., 1995b; Smit et al., 1996).

The role of receptor phosphorylation in signaling begins to be elucidated. Classically, two categories of hormone-induced biological effects are distinguished, whether hormonal stimulation leads to cell proliferation or to transcriptional activation of target genes reflecting the state of cell differentiation. Functional analysis of hGHR mutants has shown that GH-induced proliferation of the promyeloid FDC-P1 cell line does not require phosphorylated tyrosines (Wang et al., 1995) and can still be mediated by a truncated GHR containing only the 54 membrane-proximal aa of the cytoplasmic tail (Colosi et al., 1993). In the context of the PRLR, the non-phosphorylated short PRLR can induce proliferation of NIH-3T3 fibroblasts (Das and Vonderhaar, 1995), but not of FDC-P1 and BaF-3 hematopoietic cells (O'Neal and Yu-Lee, 1994). Accordingly, a non-phosphorylated mutant form of the long PRLR, truncated after 94 cytoplasmic residues, maintains the ability to transduce a proliferative signal in murine 32D cells (DaSilva et al., 1994). In contrast, transactivation of reporter genes (luciferase/chloramphenicol acetal transferase) controlled by promoters of PRL or GH target genes (Serine protease inhibitor [Spi] 2.1, β-casein) requires additional C-terminal region(s) of the receptors and, most often, phosphorylated tyrosines (Goujon et al., 1994; Hackett et al., 1995; Lebrun et al., 1995a,b; Lobie et al., 1995; VanderKuur et al., 1995b; Sotiropoulos et al., 1996). For example, the short PRLR (Lebrun et al., 1995b), as well as the non-phosphorylated Nb2 mutant lacking the C-terminal tyrosine 382 (Lebrun et al., 1995a), are unable to stimulate the β-casein promoter. Similarly, a C-terminal truncated GHR is active in Spi 2.1 transactivation, while point mutations of Tyr 469 and Tyr 516 prevent this differentiation signal (Sotiropoulos et al., 1996).

Although attempts to draw general rules from these data still remain premature, it appears that i) Box 1 is almost an absolute requirement for any PRLR/GHR signaling; ii) receptor tyrosine phosphorylation (by JAK2) is not required for proliferation of all cell types; and iii) in contrast, receptor tyrosine phosphorylation is required for most differentiation signals. However, one must keep in mind first, that results may diverge significantly depending on the cell system used and, second, that data obtained in the particular context of truncated receptors hide the presumed redundancy of some features found in full length receptors, such as the multiplicity of tyrosine residues.

Phosphatases

Since activation of several molecules implicated in PRLR/GHR transduction occurs through tyrosine phosphorylation, the involvement of tyrosine phosphatases in order to modulate or downregulate phosphorylation events in signaling cascades is not surprising. Accordingly, several recent studies pointed out a role of tyrosine phosphatases in PRLR/GHR signaling (Tourkine et al., 1995; Ali et al., 1996; Daniel et al., 1996; Sotiropoulos et al., 1996), although the mechanism by which they are activated, as well as their substrates, remains poorly documented. Due to the activating potency of tyrosine phosphorylation on signaling proteins, one would anticipate tyrosine phosphatases to negatively regulate these signals. In contrast, we have recently reported that the phosphatase PTP-1D (now renamed SHP-2 (Adachi et al., 1996)), activated by JAK2, acts as a positive regulator of PRLR-dependent induction of β-casein gene transcription (Ali et al., 1996). According to the observation that the kinetic of JAK2 dephosphorylation is retarded in cells expressing a C-terminal tail truncated form of the GHR, this region was proposed to be a site of interaction between the receptor and a phosphatase (Sotiropoulos et al., 1996).

STATs

Signal transducers and activator of transcription (STAT) are a family of latent cytoplasmic proteins of ~90–100 kDa recently identified on the basis of their involvement in cytokine receptor signaling (Darnell et al., 1994; Finidori and Kelly, 1995; Ihle and Kerr, 1995; Ihle, 1996a). The STAT gene family currently contains eight members: STAT1 (α and β), STAT2, STAT3, STAT4, STAT5a, STAT5b, STAT6 (or IL-4 STAT), and dSTAT, a STAT homologue found in *Drosophila*. STATs contain five conserved features: a DNA binding domain, a SH3-like domain, a SH2 domain, an ubiquitous tyrosine, and a C-terminal transactivating domain (from the N- to C-terminus, respectively). A consensus model of STAT activation has been proposed on the basis of data collected from studies of the different cytokine receptors (Shuai et al., 1994; Finidori and Kelly, 1995; Ihle and Kerr, 1995; Taniguchi, 1995; Ihle, 1996a). Upon hormonal activation, cytokine receptors undergo tyrosine phosphorylation by the associated Janus kinase(s) and these phosphorylated tyrosines then become binding sites for STAT SH2 domains. STATs are phosphorylated by the receptor-associated Janus kinases, dissociate from the receptor through a mechanism that remains unknown and then homo- or heterodimerize. This STAT-STAT dimerization is believed to be mediated by the interaction between the phosphotyrosine of each monomer and the SH2 domain of the other STAT molecule. Through a mechanism that is still not understood, STAT dimers then translocate to the nucleus where they interact with and activate specific DNA elements found in the promoters of cytokine target genes. Consensus DNA motifs specifically recognized by STAT complexes have been identified in these promoters. The motif termed GAS (for gamma interferon activated sequence) was defined using STAT homodimers and consists in a palindromic sequence TTC xxx GAA (Horseman and Yu-Lee, 1994; Ihle, 1996a). The specificity of the interaction between a particular STAT and a GAS motif found in a given target promoter has been proposed to depend, at least in part, on the center core nucleotide(s) (Ihle, 1996a). Although this model of STAT activation is presumably relevant, recent data suggest that the picture is probably more complicated. For example, phosphotyrosines not only of the receptors, but also of the receptor-associated Janus kinases, seem able to interact with STAT SH2 domains (Gupta et al., 1996; Smit et al., 1996; Sotiropoulos et al., 1996). Moreover, it has been shown that a single phosphotyrosyl-STAT SH2 interaction is sufficient to ensure STAT dimerization (Gupta et al., 1996). Finally, STAT transactivating properties appear, at least in some cell types, to also be regulated by serine/threonine kinases (Beadling et al., 1996; Ihle, 1996b; Ram et al., 1996).

Each cytokine receptor can only activate a limited subset of STAT proteins, and it is believed that the limitations of these combinations are, at least in part, responsible for the specificity of signals transduced by the different cytokine receptors in a given cell type (see below). Three members of the STAT family have thus far been clearly identified as transducer molecules of PRL/GH receptors: STAT1, STAT3 and STAT5.

STAT5, previously referred to as mammary gland factor (MGF), was cloned as a transcription factor typically induced by prolactin in sheep mammary gland (Gouilleux et al., 1994; Wakao et al., 1994). It was further shown that STAT5 is also activated by GH (Gouilleux et al., 1995; Wood et al., 1995; Galsgaard et al., 1996; Sotiropoulos et al., 1996) as well as by a large variety of cytokine receptors (Gouilleux et al., 1995; Mui et al., 1995; Pallard et al., 1995; Wakao et al., 1995). It has been demonstrated that the DNA binding activity of STAT5 requires phosphorylation of a single tyrosine (Tyr 694) which is mediated by JAK2, in agreement with the findings that PRLR and GHR deleted of their Box 1 (thus unable to associate/activate Janus kinases) cannot activate Stat5 (DaSilva et al., 1996; Sotiropoulos et al., 1996). The C-terminal domain of STAT5 (containing the phosphorylated tyrosine) is functionally essential since truncation after Tyr 683 leads to dominant-negative protein (Mui et al., 1996). In the current state of the art, activated STAT5 is supposed to homodimerize prior to its nuclear translocation, although its participation in heterogeneous DNA binding protein complexes has been suggested (Kazansky et al., 1996). Accordingly, STAT5a and STAT5b have also been shown to heterodimerize (Quelle et al., 1996). Activation of STAT5 by GHR appears to be dependent on receptor tyrosine phosphorylation (Sotiropoulos et al., 1996). We have also shown that mutation of the C-terminal tyrosine 382 in the Nb2 PRLR, which abolishes tyrosine phosphorylation of the receptor, prevents activation of the β-casein promoter (Lebrun et al., 1995a), suggesting that this particular residue is involved in STAT5 recruitment and/or activation. The picture remains unclear, however, since recent data suggest that a C-terminal truncated form of the Nb2 PRLR (termed G328, lacking Tyr 382) is able to stimulate STAT5 tyrosine phosphorylation (DaSilva et al., 1996). This could mean that STAT5 tyrosine phosphorylation is required, but not necessarily sufficient for activation of transcriptional activity. A possible explanation would be the requirement of serine/threonine phosphorylation, which has been recently reported for IL-2R activation of STAT5 (Beadling et al., 1996).

STAT1 and STAT3 are both activated by the PRLR and the GHR (David et al., 1994; Gronowski and Rotwein, 1994; Meyer et al., 1994; Campbell et al., 1995; Lebrun et al., 1995b; Sotiropoulos et al., 1995; DaSilva et al., 1996). STAT3, also named APRF for acute phase response protein, was cloned as an IL-6 activated transcription factor (Lütticken et al., 1994). STAT1 was first isolated as part of the interferon-stimulated gene factor 3 (ISGF3) complex (containing STAT2 and a DNA-binding protein called p48) which is typically activated by IFNs α and β (Levy et al., 1989). STAT1 homodimers can also be formed upon IFNγ activation. The receptor regions required for activation of these STATs remain poorly documented. In the context of the GHR, we have recently hypothesized that phosphotyrosine(s) of JAK2 could bind to STAT3, even if such interaction does not necessarily preclude the possible occurrence of interactions with the receptor (Smit et al., 1996; Sotiropoulos et al., 1996). Concerning the PRLR, the 93 membrane-proximal residues have been reported sufficient to activate tyrosine phosphorylation of both STAT1 and 3 (DaSilva et al., 1996). Although attempts have been made to correlate STAT binding abilities of cytokine receptors with the presence of conserved aa surrounding cytoplasmic phosphotyrosines (Stahl et al., 1995; Gerhartz et al., 1996; Hemman et al., 1996; Smit et al., 1996), identification of any consensus sequence in the PRLR and GHR for binding to STAT1, 3, and 5 has been unsuccessful.

The activation of identical STAT proteins by different cytokine receptors questions the mechanisms by which specificity of signalling pathways is achieved in response to a particular hormonal stimulation. For example, although several cytokines (EPO, GM-CSF, GH, PRL, IL-2, IL-3, IL-5) activate the DNA binding ability of STAT5 and/or transactivate the β-casein luciferase reporter gene *in vitro* (Fujii et al., 1995; Gouilleux et al., 1995; Mui et al., 1995; Pallard et al., 1995; Wakao et al., 1995; Ferrag et al., 1996; Sotiropoulos et al., 1996), it is unlikely that all these cytokines stimulate the synthesis of this milk protein *in vivo*. This suggests that different STAT combinations and/or involvement of other signal transducers direct the specificity of the final response.

Other Signaling Pathways

Although the JAK-STAT cascade is presumably the most important signaling pathway used by PRL/GH receptors, other transducing molecules are also likely involved in transduction by these receptors. Among these, one can cite pathways involving the Ras/Raf/microtubule-associated protein kinase (MAPK) cascade, IRS-1, PI-3 kinase, protein kinase C (PKC), phospholipase C (PLC)-γ or Ca^{2+} ions, in spite of the fact that these pathways remain less well documented. Signaling through MAPK involves the Shc/SOS/Grb2/Ras/Raf/MAPK cascade (Avruch et al., 1994). Activation of the MAPK pathway has been reported in different biological systems under PRL (Buckley et al., 1994; Clevenger et al., 1994; Piccoletti et al., 1994; Das and Vonderhaar, 1995; Erwin et al., 1996) and GH (Campbell et al., 1992; Sotiropoulos et al., 1994; VanderKuur et al., 1995a) stimulation. IRS-1 is phosphorylated, presumably by JAK2, upon GHR (Argetsinger et al., 1995) activation. Phosphotyrosyl residues of insulin receptor substrate-1 (IRS-1) interact with several SH2 containing proteins, among which the regulatory p85 subunit of phosphatidylinisitol (PI)-3 kinase, and it has been shown that inhibitors of this enzyme block the insulin-like effects of GH (Ridderstrale and Tornqvist, 1994). PLC generates diacylglycerol which in turn activates PKC. Activation of these enzymes by stimulated PRLR and GHR has been suggested (Buckley et al., 1988; Doglio et al., 1989), although their role in signaling remains unknown and their substrates poorly identified (Liu, 1996). Finally, PKC-independent increase of intracellular calcium has also been reported for both receptors and, at least in the GHR context, this phenomenon seems independent of JAK2 activation but requires the ~200 C-terminal residues (Ilondo et al., 1994; Billestrup et al., 1995).

PHYSIOLOGICAL ROLE OF THE RECEPTOR ISOFORMS

As described above, PRLR and GHR exist under several isoforms and it is likely that the heterogeneity of these receptors is larger than expected (Clevenger et al., 1995; Nagano et al., 1995). In addition to the membrane-bound receptors, soluble PRL or GH binding proteins have been isolated in several species (Leung et al., 1987; Postel-Vinay et al., 1991; Fuh and Wells, 1995). Their role remains unclear (Baumann, 1995). By protecting bound hormones from clearance, they have been proposed to increase the half-life of their ligands. Moreover, although binding proteins have no intracellular domain and are thus devoid of signaling potency per se, it has been proposed that ligand-mediated interaction of these binding proteins with membrane PRLR receptors (or other transducer molecules) could initiate signal transduction in cells (Lesueur et al., 1993). The physiological role of the short PRLR isoform is more intriguing. This isoform, which is unable to activate typical PRL signaling pathways (see above), is predominant in some tissues such as liver. Determining whether the physiological role of the short PRLR is restricted to stimulation of proliferation in some cell types (O'Neal and Yu-Lee, 1994; Das and Vonderhaar, 1995) or is extended to other bioactivities is a future challenge in understanding PRLR functions. In this aim, the recent cloning of an ovarian-specific p32 phosphoprotein associated with the short, but presumably not the long, PRLR (Duan et al., 1996) might be the beginning of an answer.

To better define the specific and multiple roles of lactogens, we have produced mice carrying a germline null mutation of the prolactin receptor gene (Ormandy et al., 1997). Analysis of reproduction phenotypes shows multiple abnormalities, including absence of pseudopregnancy, ovulation of premeiotic oocytes, reduced fertilization of oocytes, reduced preimplantation oocyte development, lack of embryo implantation in homozygous (-/-) females, and reduced fertility or even infertility in homozygous males. In view of the widespread distribution of PRL receptors, other phenotypes are currently being evaluated. This study establishes the prolactin receptor as a key regulator of mammalian reproduction, and provides the first total ablation model (all isoforms are knocked out) to further study the role of the prolactin receptor and its ligands. Future development of mice deleted from a single receptor isoform (soluble, membrane, long, short) will help in understanding the physiological significance of the PRL and GH receptor heterogeneity.

ACKNOWLEDGMENTS

The authors thank S. Kinet for help with the 3D figure of hPRL. V. Goffin acknowledges the EEC for fellowship support. F. Ferrag is grateful to Dr. M. Saunders and acknowledges Glaxo Company (France) for financial support.

REFERENCES

Abdel-Meguid, S.S., Shieh, H.S., Smith, W.W., Dayringer, H.E., Violand, B.N., & Bentle, L.A. (1987). Three-dimensional structure of a genetically engineered variant of porcine growth hormone. Proc. Natl. Acad. Sci. USA 84, 6434-6437.

Adachi, M., Fisher, E.H., Ihle, J., Imai, K., Jirik, F., Neel, B., Pawson, T., Shen, S.H., Thomas, M., Ullrich, A., & Zhao, Z. (1996). Mammalian SH2-containing protein tyrosine phosphatases. Cell 85, 15.

Ali, S., Pellegrini, I., Edery, M., Lesueur, L., Paly, J., Djiane, J., & Kelly, P.A. (1991). A prolactin-dependent immune cell line (Nb2) expresses a mutant form of prolactin receptor. J. Biol. Chem. 266, 20110-20117.

Ali, S., Chen, Z., Lebrun, J.J., Vogel, W., Kharitonenkov, A., Kelly, P.A., & Ullrich, A. (1996). PTP1D is a positive regulator of the prolactin signal leading to β-casein promoter activation. EMBO J. 15, 135-142.

Argetsinger, L.S., Campbell, G.S., Yang, X., Witthuhn, B.A., Silvennoinen, O., Ihle, J.N., & Carter-Su, C. (1993). Identification of JAK2 as a growth hormone receptor-associated tyrosine kinase. Cell 74, 237-244.

Argetsinger, L.S., Ilsu, G.W., Myers, M.G., Billestrup, N., White, M., & Carter-Su, C. (1995). Growth hormone, interferon-g, and leukemia inhibitory factor promoted tyrosyl phosphorylation of insulin receptor substrate-1. J. Biol. Chem. 270, 14685-14692.

Avruch, J., Zhang, X-f., & Kyriakis, J.M. (1994). Raf meets Ras: completing the framework of a signal transduction pathway. TIBS 19, 279-283.

Bass, S.H., Mulkerrin, M.G., & Wells, J.A. (1991). A systematic mutational analysis of hormone-binding determinants in the human growth hormone receptor. Proc. Natl. Acad. Sci. USA 88, 4498-4502.

Baumann, G. (1995). Growth hormone-binding protein: errant receptor or active player. Endocrinology 136, 377-378.

Baumbach, W.R., Horner, D.L., & Logan, J.S. (1989). The growth hormone-binding protein in rat serum is an alternatively spliced form of the rat growth hormone receptor. Genes Dev. 3, 1199-1205.

Baumgartner, J.W., Wells, C.A., Chen, C.M., & Waters, M.J. (1994). The role of the WSxWS motif in growth hormone receptor function. J. Biol. Chem. 269, 29094-29101.

Bazan, F. (1989). A novel family of growth factor receptors: a common binding domain in the growth hormone, prolactin, erythropoietin and IL-6 receptors, and p75-IL-2 receptor β-chain. Biochem. Biophys. Res. Commun. 164, 788-795.

Bazan, J.F. (1990). Structural design of molecular evolution of a cytokine receptor superfamily. Proc. Natl. Acad. Sci. USA 87, 6934-6938.

Beadling, C., Ng, J., Baddage, J.W., & Cantrell, D.A. (1996). Interleukin-2 activation of STAT5 requires the convergent action of tyrosine kinases and a serine/threonine kinase pathway distinct from the Raf1/ERK2 MAP kinase pathway. EMBO J. 15, 1902-1913.

Berlanga, J.J., Fresno Vara, J.A., Martin-Perez, J., & Garcia-Ruiz, J.P. (1995). Prolactin receptor is associated with c-src kinase in rat liver. Mol. Endocrinol. 9, 1461-1467.

Billestrup, N., Bouchelouche, P., Allevato, G., Ilondo, M., & Nielsen, J.H. (1995). Growth hormone receptor C-terminal domains required for growth hormone-induced intracellular free Ca^{2+} oscillations and gene transcription. Proc. Natl. Acad. Sci. USA 92, 2725-2729.

Boutin, J.M., Jolicoeur, C., Okamura, H., Gagnon, J., Edery, M., Shirota, M., Banville, D., Dusanter-Fourt, I., Djiane, J., & Kelly, P.A. (1988). Cloning and expression of the rat prolactin receptor, a member of the growth hormone/prolactin receptor gene family. Cell 53, 69-77.

Buckley, A.R., Crowe, P.D., & Russel, H. (1988). Rapid activation of a proteine kinase C in isolated rat liver nuclei by prolactin, a known hepatic mitogen. Proc. Natl. Acad. Sci. USA 85, 8649-8653.

Buckley, A.R., Rao, Y-P., Buckley, D.J., & Gout, P.W. (1994). Prolactin-induced phosphorylation and nuclear translocation of MAP Kinase in Nb2 lymphoma cells. Biochem. Biophys. Res. Commun. 204, 1158-1164.

Campbell, G.S., Pang, L., Miyasaka, T., Saltiel, A.R., & Carter-Su, C. (1992). Stimulation by growth hormone of MAP kinase activity in 3T3-F422A fibroblasts. J. Biol. Chem. 267, 6074-6080.

Campbell, G.S., Argetsinger, L.S., Ihle, J.N., Kelly, P.A., Rillema, J.A., & Carter-Su, C. (1994). Activation of JAK2 tyrosine kinase by prolactin receptors in Nb2 cells and mouse mammary gland explants. Proc. Natl. Acad. Sci. USA 91, 5232-5236.

Campbell, G.S., Meyer, D.J., Raz, R., Levy, D.E., Schwartzt, J., & Carter-Su, C. (1995). Activation of acute phase response factor (APRF)/stat3 transcription factor by growth hormone. J. Biol. Chem. 270, 3974-3979.

Caravella, J.A., Lyne, P.D., & Richards, W.C. (1996). A partial model of the erythropoietin receptor complex. Proteins 24, 394-401.

Carter-Su, C., Stubbart, J.R., Wang, X., Stred, S.E., Argetsinger, L.S., & Shafer, J.A. (1989). Phosphorylation of highly purified growth hormone receptors by a growth hormone receptor-associated tyrosine kinase. J. Biol. Chem. 264, 18654-18661.

Clarckson, T., & Wells, J.A. (1995). A hot spot of binding energy in a hormone-receptor interface. Science 267, 383-386.

Clevenger, C.V., & Medaglia, M.V. (1994). The protein tyrosine kinase p59fyn is associated with prolactin (PRL) receptor and is activated by PRL stimulation of T-lymphocytes. Mol. Endocrinol. 8, 674-681.

Clevenger, C.V., Torigoe, T., & Reed, J.C. (1994). Prolactin induces rapid phosphorylation and activation of prolactin receptor-associated RAF-1 kinase in a T-cell line. J. Biol. Chem. 269, 5559-5565.

Clevenger, C.V., Chang, W.P., Ngo, W., Pasha, T.L.M., Montone, K.T., & Tomaszewski, J.E. (1995). Expression of prolactin and prolactin receptor in human breast carcinoma. Am. J. Pathol. 146, 695-705.

Colosi, P., Wong, K., Leong, S.R., & Wood, W.I. (1993). Mutational analysis of the intracellular domain of the human growth hormone receptor. J. Biol. Chem. 268, 12617-12623.

Cunningham, B.C., & Wells, J.A. (1989). High-resolution epitope mapping in hGH-receptor interactions by alanine-scanning mutagenesis. Science 244, 1081-1084.

Cunningham, B.C., Ultsch, M., De Vos, A.M., Mulkerrin, M.G., Clauser, K.R., & Wells, J.A. (1991). Dimerization of the extracellular domain of the human growth hormone receptor by a single hormone molecule. Science 254, 821-825.

Cunningham, B.C., & Wells, J.A. (1991). Rational design of receptor-specific variants of human growth hormone. Proc. Natl. Acad. Sci. USA 88, 3407-3411.

Daniel, N., Waters, M.J., Bignon, C., & Djiane, J. (1996). Involvement of a subset of tyrosine kinases and phosphatases in regulation of the β-lactoglobulin gene promoter by prolactin. Mol. Cell. Endocrinol. 118, 25-35.

Darnell, J.E., Jr., Kerr, I.M., & Stark, G.R. (1994). Jak-STAT pathways and transcriptional activation in response to IFNs and other extracellular signaling proteins. Science 264, 1415-1421.

Das, R., & Vonderhaar, B.K. (1995). Transduction of prolactin's (PRL) growth signal through both long and short forms of the PRL receptor. Mol. Endocrinol. 9, 1750-1759.

DaSilva, L., Howard, O.M.Z., Rui, H., Kirken, R.A., & Farrar, W.L. (1994). Growth signaling and Jak2 association mediated by membrane-proximal regions of prolactin receptors. J. Biol. Chem. 269, 18267-18270.

DaSilva, L., Rui, H., Erwin, R.A., Zack Howard, O.M., Kirken, R.A., Malabarba, M.G., Hackett, R.H., Larner, A.C., & Farrar, W.L. (1996). Prolactin recruits STAT1, STAT3 and STAT5 independent of conserved receptor tyrosines TYR402, TYR476, TYR515 and TYR580. Mol. Cell. Endocrinol. 117, 131-140.

David, M., Petricoin III, E.F., Igarashi, K.-I., Feldman, G.M., Finbloom, D.S., & Larner, A.C. (1994). Prolactin activates the interferon-regulated p91 transcription factor and the Jak2 kinase by tyrosine phosphorylation. Proc. Natl. Acad. Sci. USA 91, 7174-7178.

De Vos, A.M., Ultsch, M., & Kossiakoff, A.A. (1992). Human growth hormone and extracellular domain of its receptor: crystal structure of the complex. Science 255, 306-312.

Dinerstein, H., Lago, F., Goujon, L., Ferrag, F., Esposito, N., Finidori, J., Kelly, P.A., & Postel-Vinay, M.C. (1995). The proline-rich region of the growth hormone receptor is essential for Jak2 phosphorylation, activation of cell proliferation and gene transcription. Mol. Endocrinol. 9, 1701-1707.

Doglio, A., Dani, C., Grimaldi, P., & Ailhaud, G. (1989). Growth hormone stimulates c-fos gene expression by means of protein kinase C without increasing inositol lipid turnover. Proc. Natl. Acad. Sci. USA 86, 1148-1152.

Duan, W.R., Linzer, D.I.H., & Gibori, G. (1996). Cloning and characterization of an ovarian-specific protein that associates with the short form of the prolactin receptor. J. Biol. Chem. 271, 15602-15607.

Duquesnoy, P., Sobrier, M.L., Duriez, B., Dastot, F., Buchanan, C.R., Savage, M.O., Preece, M.A., Craescu, C.T., Blouquit, Y., Goossens, M., & Amselem, S. (1994). A single amino acid substitution in the exoplasmic domain of human growth hormone (GH) receptor confers familial GH resistance (Laron syndrome) with positive GH-binding activity by abolishing receptor homodimerization. EMBO J. 13, 1386-1395.

Duriez, B., Sobrier, M.L., Duquesnoy, P., Tixier-Boichard, M., Decuypere, E., Coquerelle, G., Zeman, M., Goossens, M., & Amselem, S. (1993). A naturally occuring growth hormone receptor mutation: in vivo and in vitro evidence for the functional importance of the WS motif common to all members of the cytokine receptor superfamily. Mol. Endocrinol. 7, 806-814.

Dusanter-Fourt, I., Muller, O., Ziemiecki, A., Mayeux, P., Drucker, B., Djiane, J., Wilks, A., Harper, A.G., Fischer, S., & Gisselbrecht, S. (1994). Identification of Jak protein tyrosine kinases as signaling molecules for prolactin. Functional analysis of prolactin receptor and prolactin-erythropoietin receptor chimera expressed in lymphoid cells. EMBO J. 13, 2583-2591.

Erpel, T., & Courtneidge, S.A. (1995). *Src* family protein tyrosine kinases and cellular signal transduction pathways. Curr. Opin. Cell. Biol. 7, 176-182.

Erwin, R.A., Kirken, R.A., Malabarba, M.G., Farrar, W.L., & Rui, H. (1996). Prolactin activates ras via signaling proteins Shc, growth factor receptor bound 2, and Son of Sevenless. Endocrinology 136, 3512-3518.

Ferrag, F., Chiarenza, A., Goffin, V., & Kelly, P.A. (1996). Convergence of signaling transduced by PRL/cytokine chimeric receptors on PRL-responsive gene transcription. Mol. Endocrinol. 10, 451-460.

Finidori, J., & Kelly, P.A. (1995). Cytokine receptor signalling through two novel families of transducer molecules: Janus kinases, and signal transducers and activators of transcription. J. Endocrinol. 147, 11-23.

Frank, S.J., Yi, W., Zhao, Y., Goldsmith, J.F., Gilliland, G., Jiang, J., Sakai, I., & Kraft, A.S. (1995). Regions of the JAK2 tyrosine kinase required for coupling to the growth hormone receptor. J. Biol. Chem. 270, 14776-14785.

Fuh, G., Cunningham, B.C., Fukunaga, R., Nagata, S., Goeddel, D.V., & Wells, J.A. (1992). Rational design of potent antagonists to the human growth hormone receptor. Science 256, 1677-1679.

Fuh, G., Colosi, P., Wood, W.I., & Wells, J.A. (1993). Mechanism-based design of prolactin receptor antagonists. J. Biol. Chem. 268, 5376-5381.

Fuh, G., & Wells, J.A. (1995). Prolactin receptor antagonists that inhibit the growth of breast cancer cell lines. J. Biol. Chem. 270, 13133-13137.

Fujii, H., Nakagawa, Y., Schindler, U., Kawahara, A., Mori, H., Gouilleux, F., Groner, B., Ihle, J.N., Minami, Y., Miyazaki, T., & Taniguschi, T. (1995). Activation of STAT5 by interleukin 2 requires a carboxy-terminal region of the interleukin 2 receptor β chain but is not essential for the proliferative signal transmission. Proc. Natl. Acad. Sci. USA 92, 5482-5486.

Galsgaard, E.D., Gouilleux, F., Groner, B., Serup, P., Nielsen, J.H., & Billestrup, N. (1996). Identification of a growth hormone-responsive STAT5-binding element in the rat insulin 1 gene. Mol. Endocrinol. 10, 652-660.

Gerhartz, C., Heesel, B., Sasse, J., Hemman, U., Landgraf, C., Schneider-Mergener, J., Horn, F., Heinrich, P.C., & Graeve, L. (1996). Differential activation of acute phase response factor/STAT3 and STAT1 via the cytoplasmic domain of the interleukin 6 signal transducer gp130. I. Definition of a novel phosphotyrosine motif mediating Stat1 activation. J. Biol. Chem. 271, 12991-12998.

Ginsburg, E., & Vonderhaar, B.K. (1995). Prolactin synthesis and secretion by human breast cancer cells. Cancer Res. 55, 2591-2595.

Goffin, V., Norman, M., & Martial, J.A. (1992). Alanine-scanning mutagenesis of human prolactin: importance of the 58-74 region for bioactivity. Mol. Endocrinol. 6, 1381-1392.

Goffin, V., Struman, I., Mainfroid, V., Kinet, S., & Martial, J.A. (1994). Evidence for a second receptor binding site on human prolactin. J. Biol. Chem. 269, 32598-32606.

Goffin, V., Martial, J.A., & Summers, N.L. (1995). Use of a model to understand prolactin and growth hormone specificities. Prot. Eng. 8, 1215-1231.

Goffin, V., Kinet, S., Ferrag, F., Binart, N., Martial, J.A., & Kelly, P.A. (1996a). Antagonistic properties of human prolactin analogs that show paradoxical agonistic activity in the Nb2 bioassay. J. Biol. Chem. 271, 16573-16579.

Goffin, V., Shiverick, K.T., Kelly, P.A., & Martial, J.A. (1996b). Sequence-function relationships within the expanding family of prolactin, growth hormone, placental lactogen and related proteins in mammals. Endocr. Rev. 17, 385-410.

Gouilleux, F., Wakao, H., Mundt, M., & Groner, B. (1994). Prolactin induces phosphorylation of Tyr694 of STAT5(MGF), a prerequisite for DNA binding and induction of transcription. EMBO J. 13, 4361-4369.

Gouilleux, F., Pallard, C., Dusanter-Fourt, I., Wakao, H., Haldosen, L-A., Norstedt, G., Levy, D., & Groner, B. (1995). Prolactin, growth hormone, erythropoietin and granulocyte-macrophage colony stimulating factor induce MGF-STAT5 DNA binding activity. EMBO J. 14, 2005-2013.

Goujon, L., Allevato, G., Simonin, G., Paquereau, L., Le Cam, A., Clark, J., Nielsen, J.H., Djiane, J., Postel-Vinay, M.C., Edery, M., & Kelly, P.A. (1994). Cytoplasmic domains of the growth hormone receptor necessary for signal transduction. Proc. Natl. Acad. Sci. USA 91, 957-961.

Gourdou, I., Gabou, L., Paly, J., Kermabon, A.Y., Belair, L., & Djiane, J. (1996). Development of a constitutively active mutant form of the prolactin receptor, a member of the cytokine receptor family. Mol. Endocrinol. 10, 45-56.

Gronowski, A.M., & Rotwein, P. (1994). Rapid changes in nuclear protein tyrosine phosphorylation after Growth Hormone treatment *in vivo* . J. Biol. Chem. 269, 7874-7878.

Gupta, S., Yan, H., Wong, L.H., Ralph, S., Krolewski, J., & Schindler, C. (1996). The SH2 domains of STAT1 and STAT2 mediate multiple interactions in the transduction of INF-α signals. EMBO J. 15, 1075-1084.

Hackett, R.H., Wang, Y.D., & Larner, A.C. (1995). Mapping of the cytoplasmic domain of the human growth hormone receptor required for the activation of JAK2 and STAT proteins. J. Biol. Chem. 270, 21326-21330.

Hemman, U., Gerhartz, C., Heesel, B., Sasse, J., Kurapkat, G., Grötzinger, J., Wollmer, A., Zhong, Z., Darnell, J.E.Jr, Graeve, L., Heinrich, P.C., & Horn, F. (1996). Differential activation of acute phase response factor/STAT3 and STAT1 via the cytoplasmic domain of the interleukin 6 signal transducer gp130. II. Src homology SH2 domains define specificity of stat factor activation. J. Biol. Chem. 271, 12999-13007.

Hilton, D.J., Watowich, S.S., Katz, L., & Lodish, H.F. (1996). Saturation mutagenesis of the WSXWS motif of the erythropoietin receptor. J. Biol. Chem. 271, 4699-4708.

Hooper, K.P., Padmanabhan, R., & Ebner, K.E. (1993). Expression of the extracellular domain of the rat liver prolactin receptor and its interaction with ovine prolactin. J. Biol. Chem. 268, 22347-22352.

Horseman, N.D., & Yu-Lee, L.Y. (1994). Transcriptional regulation by the helix bundle peptide hormones: growth hormone, prolactin, and hematopoietic cytokines. Endocr. Rev. 15, 627-649.

Ihle, J.N. (1994). Signaling by the cytokine receptor superfamily. Just another kinase story. Trends Endocrinol. Metab. 5, 137-143.

Ihle, J.N., & Kerr, I.M. (1995). Jaks and STATs in signaling by the cytokine receptor superfamily. TIG 11, 69-74.

Ihle, J.N. (1996a). STATs: signal transducers and activators of transcription. Cell 84, 331-334.

Ihle, J.N. (1996b). STATs and MAPKs: obligate or opportunistic partners in signaling. Bioessays 18 (2), 95-98.

Ilondo, M.M., De Meyts, P., & Bouchelouche, P. (1994). Human growth hormone increases cytosolic free calcium in cultured human IM-9 lymphocytes: a novel mechanism of growth hormone transmembrane signalling. Biochem. Biophys. Res. Commun. 202, 391-397.

Kazansky, A.V., Raught, B., Lindsey, S.M., Wang, Y.F., & Rosen, J.M. (1996). Regulation of mammary gland factor/STAT5a during mammary gland development. Mol. Endocrinol. 9, 1598-1609.

Kelly, P.A., Djiane, J., Postel-Vinay, M.C., & Edery, M. (1991). The prolactin/growth hormone receptor family. Endocr. Rev. 12, 235-251.

Kinet, S., Goffin, V., Mainfroid, V., & Martial, J.A. (1996). Characterization of lactogen receptor binding site 1 of human prolactin. J. Biol. Chem. 271, 14353-14360.

Kishimoto, T., Taga, T., & Akira, S. (1994). Cytokine signal transduction. Cell 76, 253-262.

Kitamura, T., Ogorochi, T., & Miyajima, A. (1994). Multimeric cytokine receptors. Trends Endocrinol. Metab. 5, 8-14.

Kossiakoff, A.A., Somers, W., Ultsch, M., Andow, K., Muller, Y.A., & De Vos, A.M. (1994). Comparison of the intermediate complexes of human growth hormone bound to the human growth hormone and prolactin receptors. Protein Sci. 3, 1697-1705.

Lebrun, J.J., Ali, S., Sofer, L., Ullrich, A., & Kelly, P.A. (1994). Prolactin induced proliferation of Nb2 cells involves tyrosine phosphorylation of the prolactin receptor and its associated tyrosine kinase. J. Biol. Chem. 269, 14021-14026.

Lebrun, J.J., Ali, S., Goffin, V., Ullrich, A., & Kelly, P.A. (1995a). A single phosphotyrosine residue of the prolactin receptor is responsible for activation of gene transcription. Proc. Natl. Acad. Sci. USA 92, 4031-4035.

Lebrun, J.J., Ali, S., Ullrich, A., & Kelly, P.A. (1995b). Proline-rich sequence-mediated JAK2 association to the prolactin receptor is required but not sufficient for signal transduction. J. Biol. Chem. 270, 10664-10670.

Lesueur, L., Edery, M., Paly, J., Kelly, P.A., & Djiane, J. (1993). Roles of the extracellular and cytoplasmic domains of the prolactin receptor in signal transduction to milk protein genes. Mol. Endocrinol. 7, 1178-1184.

Leung, D.W., Spencer, S.A., Chachianes, G., Hammonds, R.G., Collins, C., Henzel, W.J., Barnard, R., Waters, M.J., & Wood, W.I. (1987). Growth hormone receptor and serum binding protein: purification, cloning, and expression. Nature 330, 537-543.

Levy, D.E., Kessler, D.S., Pine, R.I., & Darnell, J.E. (1989). Cytoplasmic activation of ISGF3, the positive regulator of interferon-α stimulated transcription, reconstitued in vitro. Genes Dev. 3, 1362-1372.

Liu, J. (1996). Protein kinase C and its substrates. Mol. Cell. Endocrinol. 116, 1-29.

Livnah, O., Stura, E.A., Johnson, D.L., Middleton, S.A., Mulcahy, L.S., Wrighton, N.C., Dower, W.J., Jolliffe, L.K., & Wilson, I.A. (1996). Functional mimicry of a protein hormone by a peptide agonist: the EPO receptor complex at 2.8 A. Science 273, 464-471.

Lobie, P.E., Allevato, G., Nielsen, J.H., Norstedt, G., & Billestrup, N. (1995). Requirement of tyrosine residues 333 and 338 of the growth hormone (GH) receptor for selected GH-stimulated function. J. Biol. Chem. 270, 21745-21750.

Lütticken, C., Wegenka, U.M., Yuan, J., Buschmann, J., Schindler, C., Ziemiecki, ., Harpur, A.G., Wilks, A.F., Yasakawa, K., Taga, T., Kishimoto, T., Barbieri, G., Pellegrini, S., Sendtner, M., Heinrich, P.C., & Horn, F. (1994). Association of transcription factor APRF and protein kinase Jak1 with the interleukin-6 signal transducer gp130. Science 263, 89-91.

Meyer, D.J., Campbell, G.S., Cochran, B.H., Argetsinger, L.S., Larner, A.C., Finbloom, D.S., Carter-Su, C., & Schwartz, J. (1994). Growth hormone induces a DNA binding factor related to the interferon-stimulated 91-kDa transcription factor. J. Biol. Chem. 269, 4701-4704.

Miyazaki, T., Maruyama, M., Yamada, G., Hatakeyama, M., & Taniguchi, T. (1991). The intergity of the conserved WS motif common to IL-2 and other cytokine receptors is essential for ligand binding and signal transduction. EMBO J. 10, 3191-3197.

Mui, A.L-F., Wakao, H., O'Farrell, A-M., Harada, N., & Miyajima, A. (1995). Interleukin-3, granulocyte-macrophage colony stimulating factor and interleukin-5 transduce signals through two STAT5 homologs. EMBO J. 14, 1166-1175.

Mui, A.L., Wakao, H., Kinoshita, T., Kitamura, T., & Miyajima, A. (1996). Suppression of interleukin-3-induced gene expression by a C-terminal truncated Stat5: role of Stat5 in proliferation. EMBO J. 15, 2425-2433.

Murakami, M., Narazaki, M., Hibi, M., Yawata, H., Yazukawa, K., Hamaguchi, M., Taga, T., & Kishimoto, T. (1991). Critical cytoplasmic region of the IL-6 signal transducer, gp 130, is conserved in the cytokine receptor family. Proc. Natl. Acad. Sci. USA 88, 11349-11353.

Nagano, M., Chastre, E., Choquet, A., Bara, J., Gespach, C., & Kelly, P.A. (1995). Expression of prolactin and growth hormone receptor genes and their isoforms in the gastrointestinal tract. Am. J. Physiol. 268, G431-G442.

O'Neal, K.D., & Yu-Lee, L.Y. (1994). Differential signal transduction of the short, Nb2 and long prolactin receptors. J. Biol. Chem. 269, 26076-26082.

O'Neal, K.D., Chari, M.V., McDonald, C.H., Cook, R.G., Yu-Lee, L.Y., Morrisett, J.D., & Shearer, W.T. (1996). Multiple cis-trans conformers of the prolactin receptor proline-rich motif (PRM) peptide detected by reverse-phase HPLC, CD, and NMR. Biochem. J. 315, 833-844.

Ormandy, C.J., Camus, A., Barra, J., Damotte, D., Lucas, B.K., Buteau, H., Edery, M., Brousse, N., Babinet, C., Binart, N., & Kelly, P.A. (1997). Null mutation of the prolactin receptor gene produces multiple reproductive defects in the mouse. Genes Dev. 11, 167-178.

Pallard, C., Gouilleux, F., Charon, M., Groner, B., Gisselbrecht, S., & Dusanter-Fourt, I. (1995). Interleukin-3, erythropoietin and prolactin activate a STAT5 like factor in lymphoid cells. J. Biol. Chem. 270, 15942-15945.

Piccoletti, R., Maroni, P., Bendinelli, P., & Bernelli-Zazzera, A. (1994). Rapid stimulation of mitogen-activated protein kinase of rat liver by prolactin. Biochem. J. 303, 429-433.

Postel-Vinay, M.C., Belair, L., Kayser, C., Kelly, P.A., & Djiane, J. (1991). Identification of prolactin and growth hormone binding proteins in milk. Proc. Natl. Acad. Sci. USA 88, 6687-6690.

Quelle, D.E., Quelle, F.W., & Wojchowski, D.M. (1992). Mutations in the WSAWSE and cytosolic domains of the erythropoietin receptor affect signal transduction and ligand binding and internalization. Mol. Cell. Biol. 12, 4553-4561.

Quelle, F.W., Wang, D., Nosaka, T., Thierfelder, W.E., Stravopodis, D., Weinstein, Y., & Ihle, J.N. (1996). Erythropoietin induces activation of STAT5 through association with specific tyrosines on the receptor that are not required for a mitogenic response. Mol. Cell. Biol. 16, 1622-1631.

Ram, P.A., Park, S., Choi, H.K., & Waxman, D.J. (1996). Growth hormone activation of STAT1, STAT3, and STAT5 in rat liver. J. Biol. Chem. 271, 5929-5940.

Ren, R., Mayer, B.J., Cichetti, P., & Baltimore, D. (1993). Identification of ten amino acid prolin-rich SH3 binding site. Science 259, 1157-1161.

Ridderstrale, M., & Tornqvist, H. (1994). PI-3 kinase inhibitor wortmannin blocks the insulin-like effects of growth hormone in isolated rat adipocytes. Biochem. Biophys. Res. Commun. 203, 306-310.

Rozakis-Adcock, M., & Kelly, P.A. (1991). Mutational analysis of the ligand binding domain of the prolactin receptor. J. Biol. Chem. 266, 16472-16477.

Rozakis-Adcock, M., & Kelly, P.A. (1992). Identification of ligand binding determinants of the prolactin receptor. J. Biol. Chem. 267, 7428-7433.

Rui, H., Djeu, J.Y., Evans, G.A., Kelly, P.A., & Farrar, W.L. (1992). Prolactin receptor triggering: evidence for rapid tyrosine kinase activation. J. Biol. Chem. 267, 24076-24081.

Rui, H., Kirken, R.A., & Farrar, W. (1994). Activation of receptor-associated tyrosine kinase JAK2 by prolactin. J. Biol. Chem. 269, 5364-5368.

Shirota, M., Banville, D., Ali, S., Jolicoeur, C., Boutin, J-M., Edery, M., Djiane, J., & Kelly, P.A. (1990). Expression of two forms of prolactin receptor in rat ovary and liver. Mol. Endocrinol. 4, 1136-1142.

Shuai, K., Horvath, C.M., Tsai Huang, L.H., Qureshi, S.A., Cowburn, D., & Darnell, J.E.Jr (1994). Interferon activation of the transcription factor STAT91 involves dimerization through SH2-phosphotyrosyl peptide interactions. Cell 76, 821-828.

Smit, L.S., Meyer, D.J., Billestrup, N., Norstedt, G., Schwartz, J., & Carter-Su, C. (1996). The role of the growth hormone (GH) receptor and JAK1 and JAK2 kinases in the activation of STATs 1, 3, and 5 by GH. Mol. Endocrinol. 10, 519-533.

Somers, W., Ultsch, M., De Vos, A.M., & Kossiakoff, A.A. (1994). The X-ray structure of the growth hormone-prolactin receptor complex: receptor binding specificity developed through conformational variability. Nature 372, 478-481.

Sotiropoulos, A., Goujon, L., Simonin, G., Kelly, P.A., Postel-Vinay, M.C., & Finidori, J. (1993). Evidence for generation of the growth hormone-binding protein through proteolysis of the growth hormone membrane receptor. Endocrinology 132, 1863-1865.

Sotiropoulos, A., Perrot-Applanat, M., Dinerstein, H., Pallier, A., Postel-Vinay, M.C., Finidori, J., & Kelly, P.A. (1994). Distinct cytoplasmic regions of the growth hormone receptor are required for activation of JAK2, mitogen-activated protein kinase, and transcription. Endocrinology 135, 1292-1298.

Sotiropoulos, A., Moutoussamy, S., Binart, N., Kelly, P.A., & Finidori, J. (1995). The membrane proximal region of the cytoplasmic domain of the growth hormone receptor is involved in the activation of STAT3. FEBS Lett. 369, 169-172.

Sotiropoulos, A., Moutoussamy, S., Renaudie, F., Clauss, M., Kayser, C., Gouilleux, F., Kelly, P.A., & Finidori, J. (1996). Differential activation of STAT3 and STAT5 by distinct regions of the growth hormone receptor. Mol. Endocrinol. 10, 998-1009.

Stahl, N., & Yancopoulos, G.D. (1993). The alphas, betas and kinases of cytokine receptor complexes. Cell 74, 587-590.

Stahl, N., Farruggella, T.J., Boulton, T.G., Zhong, Z., Darnell, J.E., Jr., & Yancopoulos, G.D. (1995). Choice of STATs and other substrates specified by modular tyrosine-based motifs in cytokine receptors. Science 267, 1349-1353.

Sun, X.J., Pons, S., Asano, T., Myers, M.G., Glasheen, E., & White, M.F. (1996). The Fyn tyrosine kinase binds IRS-1 and forms a distinct signaling complex during insulin stimulation. J. Biol. Chem. 271, 10583-10587.

Taga, T., & Kishimoto, T. (1993). Cytokine receptors and signal transduction. FASEB J. 7, 3387-3396.

Taniguchi, T. (1995). Cytokine signaling through nonreceptor protein tyrosine kinases. Science 268, 251-255.

Tourkine, N., Schindler, C., Larose, M., & Houdebine, L.M. (1995). Activation of Stat factors by prolactin, interferon-γ, growth hormones, and a tyrosine phosphatase inhibitor in rabbit primary mammary epithelial cells. J. Biol. Chem. 270, 20952-20961.

VanderKuur, J., Allevato, G., Billestrup, N., Norstedt, G., & Carter-Su, C. (1995a). Growth hormone -promoted tyrosyl phosphorylation of SHC proteins and SHC association with Grb2. J. Biol. Chem. 270, 7587-7593.

VanderKuur, J.A., Wang, X., Zhang, L., Allevato, G., Billestrup, N., & Carter-Su, C. (1995b). Growth hormone-dependent phosphorylation of tyrosine 333 and/or 338 of the growth hormone receptor. J. Biol. Chem. 270, 21738-21744.

Wakao, H., Gouilleux, F., & Groner, B. (1994). Mammary gland factor (MGF) is a novel member of the cytokine regulated transcription factor gene family and confers the prolactin response. EMBO J. 13, 2182-2191.

Wakao, H., Harada, N., Kitamura, T., Mui, A.L., & Miyajima, A. (1995). Interleukin-2 and erythropoietin activate STAT5/MGF via distinct pathways. EMBO J. 14, 2527-2535.

Walter, M.R., Windsor, W.T., Nagabhusgan, T.L., Lundell, D.J., Lunn, C.A., Zauodny, P.J., & Narula, S.K. (1995). Crystal structure of a complex between interferon-γ and its soluble high-affinity receptor. Nature 376, 230-235.

Wang, Y.D., Wong, K., & Wood, W.I. (1995). Intracellular tyrosine residues of the human growth hormone receptor are not required for the signaling of proliferation or JAK-STAT activation . J. Biol. Chem. 270, 7021-7024.

Wang, Y.D., & Wood, W.I. (1995). Amino acids of the human growth hormone receptor that are required for proliferation and JAK-STAT signaling. Mol. Endocrinol. 9, 303-311.

Wells, J.A. (1994). Structural and functional basis for hormone binding and receptor oligomerization. Curr. Opin. Cell. Biol. 6, 163-173.

Wells, J.A. (1996). Binding in the growth hormone receptor complex. Proc. Natl. Acad. Sci. USA 93, 1-6.

Wood, T.J.J., Sliva, D., Lobie, P.E., Pircher, T., Gouilleux, F., Wakao, H., Gustafsson, J.A., Groner, B., Norstedt, G., & Haldosen, L.A. (1995). Mediation of growth hormone-dependent transcriptional activation by mammary gland factor/STAT5. J. Biol. Chem. 270, 9448-9453.

Chapter 2

Molecular Aspects of Growth Hormone Action

MICHAEL J. THOMAS and PETER ROTWEIN

Advances in Molecular and Cellular Endocrinology
Volume 2, pages 35-57.
Copyright © 1998 by JAI Press Inc.
All rights of reproduction in any form reserved.
ISBN: 0-7623-0292-5

INTRODUCTION

Growth hormone (GH), a 191 amino acid polypeptide secreted by the adenohypophysis, is a major regulator of somatic growth and metabolic functions. This review will summarize recent advances in the biology of GH action with a focus on GH receptor (GH-R) structure and function, receptor-mediated signal transduction, and subsequent intracellular effects. Regulation of GH gene expression and secretion have been the subject of many recent review articles, and will not be considered here.

GROWTH HORMONE: STRUCTURE

GH is a member of the helix-bundle peptide hormone family, a broad class of polypeptides that share a common three-dimensional structure, consisting of four antiparallel α helices (de Vos et al., 1992). In GH, these segments are arranged such that a long non-helical loop connects helices 1 and 2, a short region connects helices 2 and 3, and a long loop connects helices 3 and 4 (this is called an up-up-down-down configuration). There are two disulfide bonds, one between C^{53} in the loop between helix 1 and 2, and C^{165} of helix 4; and the other near the COOH-terminus between C^{182} and C^{189}. GH interacts with its receptor through two distinct interfaces on the surface of the GH molecule (Cunningham et al., 1991; de Vos et al., 1992; Clackson and Wells, 1995). Site 1 has an slightly concave interface of ~1,230 $Å^2$ and is composed of residues on the exposed surface of helix 4, along with residues in helix 1 and the interconnecting region between helix 1 and 2. Site 2 has a slightly smaller, flat interface of ~900 $Å^2$, and is composed of exposed residues of helices 1 and 3, and the NH_2-terminus. The K_d of GH for the GH-R has been estimated to be ~0.3 nM (Leung et al., 1987), but varies slightly according to cell type and species.

Functional domains of the GH molecule have been difficult to dissect. The integrity of Site 1 appears to be required for proliferative activity in FDC-P1 cells (Rowlinson et al., 1995). By contrast, Site 2 has been shown to be important for conferring growth-promoting activity in IM-9 cells and in transgenic mice (Chen et al., 1994, 1995). Mutational analyses of the amphiphilic region of the 3rd α helix revealed that glycine[119] in bovine GH (corresponding to glycine[120] in the human protein) is important in promoting somatic growth in transgenic mice. The significance of this residue cannot be explained entirely by perturbations in Site 2, possibly suggesting a binding interaction with another region of the GH-R (Chen et al., 1995).

In the circulation, GH is bound to the GH binding protein (GHBP), which consists of the extracellular domain of the GH-R (de Vos et al., 1992). In humans and mice, the GHBP arises from proteolytic cleavage of the cell surface GH-R (Kelly et al., 1993); however in rats, it is synthesized from an alternatively spliced GH-R mRNA and directly secreted (Baumbach et al., 1989).

GROWTH HORMONE RECEPTOR

Structure and Expression

The GH-R was cloned in 1987, using amino acid sequence information obtained from the purified rabbit protein (Leung et al., 1987). The human GH-R cDNA has an open reading frame of 638 amino acids, including an 18 amino acid signal peptide; the mature form of the receptor is 620 amino acids in length. Receptors from other species range in length from 592 residues in chicken (Burnside et al., 1991) to 626 residues in mouse (Smith et al., 1988). The 87 kb human GH-R gene is located on chromosome 5, and is composed of 9 coding exons, numbered 2–10 (Godowski et al., 1989). There are several alternative 5' exons that encode the 5' untranslated region of the GH-R mRNA. The 5' end of the mouse GH-R gene has been mapped, and a putative promoter identified (Menon et al., 1995). In the human GH-R gene, exon 2 encodes the signal peptide; exons 3–7, the extracellular portion of the GH-R; exon 8, the transmembrane region; and exons 9 and 10 encode the cytoplasmic portion of the protein and the 3' untranslated region of the mRNA.

The GH-R is expressed in many tissues. In rats, GH-R mRNA has been detected in liver, heart, kidney, intestine, skeletal muscle, pancreas, brain, and testis (Mathews et al., 1989). GH-R mRNA expression in rats is low at birth, but rises sharply at two to three weeks of age, reaching a plateau several weeks later. In chickens, hepatic GH-R mRNA appears on embryonic day 15, peaks by days 17–19, and declines on day 21, the day of hatching (Burnside and Cogburn, 1992). Approximately two weeks after birth, chicken hepatic GH-R mRNA levels begin to increase again, reaching a peak as the animal reaches mature body weight.

The cloned prolactin receptor (PRL-R) is ~30% identical to the GH-R, and both receptors are members of the cytokine receptor superfamily

(Kelly et al., 1993). This group is characterized by a single membrane spanning domain, defined sequence homologies in the extracellular domain and in the juxtamembrane intracellular region, and the absence of protein kinase activity (Kishimoto et al., 1994). The cytokine receptor superfamily has been divided into four classes (Taniguchi, 1995). Class I cytokine receptors, which include the GH-R and PRL-R, possess three pairs of disulfide bonds, and a signature motif of WSXWS in the extracellular region, adjacent to the cell membrane, although this motif is minimally conserved in the GH-R as YGEFS (de Vos et al., 1992). There is now increasing evidence that the cytokine receptors also share common components in their signal transduction pathways. Prominent among these are the cytoplasmic protein tyrosine kinases of the Janus kinase (JAK) family ((Schindler and Darnell, 1995), see below).

Ligand Binding and Dimerization

X-ray crystallographic analysis has revealed that a single molecule of GH is bound to two GH-Rs (Cunningham et al., 1991; de Vos et al., 1992), and it has been argued that dimerization of the GH-R by GH is a prerequisite for receptor activation (Ilondo et al., 1994). Because the WSXWS extracellular motif is conserved in the class I cytokine receptor superfamily, this region has been postulated to play a role in both ligand binding and signal transduction. This hypothesis was tested in the GH-R by making alanine substitutions at each site within the YGEFS sequence, and stably transfecting mutant receptors into Chinease hamster ovary (CHO) cells (Baumgartner et al., 1994). Separate mutations at Y^{222} and S^{226} diminished ligand binding and abolished signal transduction, as measured both by stimulation of protein synthesis and by transcriptional activation of a transiently transfected c-*fos* promoter reporter gene. Mutations at G^{223}, E^{224}, and F^{225} did not alter receptor function. Mutation of the YGEFS motif to the consensus WSEWS sequence produced a GH-R that was functionally similar to the wild type receptor. Although these studies do not prove that the WSXWS motif and its variant in the GH-R are involved in receptor dimerization, they do illustrate its importance for normal receptor function.

A cohort of individuals with GH insensitivity syndrome and short stature (Laron-type dwarfism) have provided a clinical demonstration of the importance of specific interactions between GH and the GH-R in mediating signal transduction pathways required for normal somatic growth.

Several of these individuals have mutations in the extracellular domain of the receptor associated with diminished hormone binding (Rosenfeld et al., 1994); however, some patients have mutations that appear to impair GH-R dimerization (Duquesnoy et al., 1994), and others harbor intracellular mutations that may alter receptor-mediated signal transduction (Kou et al., 1993; Laron et al., 1993).

GROWTH HORMONE RECEPTOR SIGNALING

Activation of JAK2

Since the discovery that interferons -α and -γ activate a family of non-receptor tyrosine kinases known as the JAKs, a number of studies have suggested a broad role for these proteins in signal transduction by the cytokine receptors (Darnell et al., 1994). There are currently four mammalian JAK: JAK1, JAK2, JAK3, and TYK2, ranging in molecular weight from 125 to 135 kDa (Schindler and Darnell, 1995). These proteins have a similar structural organization consisting of an NH_2-terminal domain that appears to interact with the cytoplasmic domain of the cytokine receptor superfamily, a COOH-terminal domain that possesses kinase activity, and a pseudokinase domain. The demonstration that GH treament activated JAK2 in 3T3-F442A fibroblasts provided the first direct connection of a GH-regulated signal transduction pathway to an intracellular protein tyrosine kinase (Argetsinger et al., 1993). This initial observation was soon confirmed in IM-9 lymphocytes (Frank et al., 1994), but it remains unresolved whether the GH-R is closely associated with JAK2 in the absence of ligand stimulation, or whether GH binding to the extracellular portion of the receptor involves a conformational change that triggers binding of its intracellular region to JAK2.

Deletion/mutation analyses of the GH-R (and other receptors that act via the JAK family) have defined a highly conserved, membrane-proximal region denoted as Box 1, which is necessary for JAK2 binding and tyrosine phosphorylation (VanderKuur et al., 1994). The 15 amino acids comprising Box 1 are rich in prolines and other hydrophobic residues. Upon ligand binding, truncated GH-R mutants containing Box 1 were able to stimulate JAK2 tyrosine phosphorylation, but lacked the ability to mediate several other actions of GH, such as activation of gene expression (Goujon et al., 1994; Sotiropoulos et al., 1994). These results indicate that domains outside of Box 1 are required for certain aspects of GH-R signal transduction. The importance of Box 1 in GH-R action has been demonstrated in studies employing site-directed mutagenesis. Substitution of all four prolines to alanines disrupted GH-inducible JAK2 phosphorylation, while partial activity was maintained when single prolines are altered. Mutated GH-Rs that lack the ability to phosphorylate JAK2 were unable to support mitogenic proliferation, induce DNA-protein interactions, or activate gene expression in stably tranfected cells (Dinerstein et al., 1995; Wang and Wood, 1995). Currently, no GH-stimulated biological effect has been shown to occur independently of JAK2 phosphorylation.

The role of an adjacent region in the intracytoplasmic part of the GH-R, termed Box 2, has not been defined. In some cytokine receptors (e.g., gp130, IL-2Rβ, G-CSF, and Epo receptors), this segment is required for ligand-stimulated mitogenesis. The PRL-R appears to require both Box 1 and Box 2 for JAK2 phosphorylation (DaSilva et al., 1994). By contrast, FDC-P1 cells transfected with a truncated human GH-R mutant containing only Box 1 were capable of GH-stimulated mitogenesis, although this effect could be augmented by the inclusion of Box 2 (Colosi et al., 1993) A point mutation of F^{346} to A (within Box 2) disrupted rat GH-R internalization, but did not block downstream effects of GH signal transduction such as gene activation (Allevato et al., 1995).

The role of tyrosine phosphorylation of the GH-R has been studied as a possible modulator/mediator of signal transduction. The GH-R becomes rapidly phosphorylated after GH binding, but a point mutation of Y^{332} to F in a truncated human GH-R containing the proximal 54 residues of the intracytoplasmic domain had no effect on ligand-stimulated proliferation (Wang et al., 1995). This same mutant receptor also maintained GH-inducible JAK2 tyrosine phosphorylation and DNA binding to a GAS element (Hackett et al., 1995). GH-stimulated tyrosine phosphorylation was diminished in a full-length rat GH-R containing mutations of Y^{333} and Y^{338} to phenylalanine, although these alterations did not perturb receptor-mediated activation of JAK2 and stimulation of tyrosine phosphorylation of other cellular proteins (VanderKuur et al., 1995). This same mutated GH-R was capable of normal high-affinity ligand binding, and ligand-activated receptor internalization, receptor downregulation, microtubule-associated protein kinase activation, and induction of reporter gene expression, but lacked the ability to stimulate lipogenesis and protein synthesis (Lobie et al., 1995). Although the function of tyrosine phosphorylation of the GH-R remains unclear, Y^{333} and Y^{338} may serve accessory roles as binding sites for some of the src homology 2 (SH_2) domain-containing proteins involved with GH-R signal transduction.

Although mutational analysis of JAK2 has not identified sites of tyrosine phosphorylation, deletion of the NH_2-terminal domain of the protein (Δ aa 2–239) impaired receptor function, suggesting that this region is required for JAK2 to associate with the GH-R (Frank et al., 1995). Similar deletional analyses have shown that no single part of the NH_2-terminal domain confers the ability of JAK2 to bind GH-R, implying that multiple regions must be present in order for this interaction to occur (Tanner et al., 1995).

Deletions of the COOH-terminal end of JAK2 (Δ aa 1,000–1,129) also impair GH-R signaling, probably by abrogating kinase activity (Frank et al., 1995). A GH-R/JAK2 chimera, composed of the extracellular and transmembrane domains of the GH-R (but lacking Box 1), fused to the COOH-terminal kinase domain of JAK2 (aa 753–1,129) could stimulate c-*fos* promoter-reporter gene expression in response to GH (Frank et al., 1995). Taken in sum, these findings suggest that the NH_2-terminal end of JAK2 is required for association with the cytoplasmic domain of the GH-R, and that the COOH-terminal kinase domain of JAK2 may mediate receptor-activated intracellular signaling.

Phosphorylation and Activation of Intracellular Proteins

Several intracellular proteins become phosphorylated in response to GH-R activation. Acute administration of GH to hypophysectomized rats induced tyrosine phosphorylation of no less than eight different nuclear proteins, including extracellular signal-regulated kinases (ERKs, also known as mitogen activated protein (MAP) kinases) 1 and 2, and signal transducers and activators of transcription (STATs) 1, 3, and 5; and these five proteins were translocated from the cytoplasm to the nucleus after GH treatment (Gronowski and Rotwein, 1994; Gronowski et al., 1995; Waxman et al., 1995). In 3T3-F442A cells, at least 13 proteins from whole cell lysates showed increased tyrosyl phosphorylation in response to GH (Campbell et al., 1993). The segments of the GH-R that are required for mediating phosphorylation of intracellular target proteins have been studied in CHO cells stably transfected with truncated rat GH-R mutants (VanderKuur et al., 1994). Deletion or mutation of Box 1 ablated intracellular tyrosine phosphorylation of at least four proteins:

p121 (JAK2); p97 (possibly STAT1 or 5); and p42 and p39 (ERKs 1 and 2). Of these four proteins, only p97 did not display GH-inducible phosphorylation when a truncated GH-R mutant (aa 1–454) was studied (VanderKuur et al., 1994). Therefore, different segments of the intracellular region of the GH-R may be required for phosphorylation of different proteins. Similar results were observed in mouse L cells transfected with the porcine GH-R (Wang et al., 1995).

In addition to inducing phosphorylation of ERKs 1 and 2, the ligand-activated GH-R could stimulate ERK activity (Campbell et al., 1992; Winston and Bertics, 1992). Initally, protein kinase C was postulated to mediate this activation, based on the inhibitory effects of staurosporine or phorbol ester pretreatment on ERK activity, but subsequent studies demonstrated a more direct link to the GH-R (Campbell et al., 1993). Transfection of mutated rat receptors into CHO cells showed that truncations of the cytoplasmic domain to aa 454 maintained the ability to stimulate ERKs, but additional truncations (to aa 380 or 294), or Box 1 mutations/deletions led to the loss of ERK activation, suggesting that Box 1 and perhaps adjacent regions may be required for this pathway (VanderKuur et al., 1994). However, other studies indicated that a truncation to aa 317 of the rabbit GH-R (containing Box 1), did not limit ERK activation when compared with the full-length wild-type receptor (Sotiropoulos et al., 1994). Figure 1 summarizes the structure-function studies discussed in this section.

Activation of STATs

The observation that novel transcription factors were induced by interferons -α and -γ led to the identification and characterization of the first two members of the STAT family of transcription factors (Darnell et al., 1994). Subsequently, different cytokine receptors were found to activate various STATs in response to ligand binding. To date, six STATs have been characterized and cloned (Ihle, 1996). These proteins comprise a related family that share ~28 to 40% sequence identity in the first ~700 amino acids, but differ widely at the COOH-terminus. They share a similar structural organization, including a DNA-binding domain (Horvath et al., 1995), a region mediating transactivation and dimerization containing SH_2 domains, and several sites of potential tyrosine phosphorylation. All STATs can form homodimers and several can form heterodimers with other STATs (Ihle, 1996). In addition, STAT1 and 3 can form a hetero-

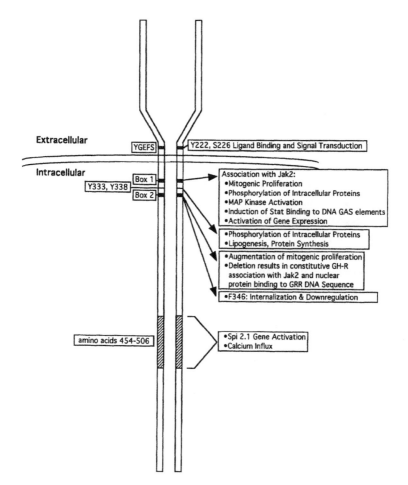

Figure 1. Schematic diagram of the structure-function relationships of the GH-R. Domains are labeled to the left of the rat GH-R dimer, and the functions based on deletion/mutation analyses are summarized to the right.

meric complex with the non-STAT protein, p48 (Veals et al., 1992; Darnell et al., 1994). Transactivation of STATs is stimulated by tyrosine phosphorylation, which confers the ability to bind to specific sequences of DNA known as g-activated sequences (GAS). These are *cis*-acting elements, consisting minimally of a core consensus sequence of $TT(N)_5AA$, that are found in the promoters of genes regulated by different cytokines. It has not been established conclusively which kinase(s) stimulate tyrosyl phosphorylation of STATs, but the JAKs are likely candidates. Following tyrosine phosphorylation, the STATs undergo translocation from the cytoplasm to the nucleus, but it is not clear whether the ability to bind DNA sequences is conferred solely by tyrosine phosphorylation, since several of these proteins also undergo serine phosphorylation, which appears to be regulated through a MAP kinase pathway (David et al., 1995; Wen et al., 1995; Zhang et al., 1995).

As observed with many other cytokine receptors, ligand-mediated stimulation of GH-R induced phosphorylation and activation of several STATs. Current studies have shown that GH treatment leads to the tyrosine phosphorylation of STAT1 (Finbloom et al., 1994; Gronowski and Rotwein, 1994; Meyer et al., 1994; Silva et al., 1994), STAT3 (Campbell et al., 1992; Gronowski et al., 1995), and STAT5 (Tourkine et al., 1995; Waxman et al., 1995; Wood et al., 1995; Gronowski et al., 1996) and several of their splicing variants. Mutations of Box 1 of the GH-R impaired the GH-inducible binding of STAT1 to a DNA sequence derived from the c-*fos* promoter (SIE). A similar effect was noted with truncation mutants of the GH-R. Nuclear protein binding to a γ-response region (GRR) DNA probe was blunted, although Box 2 was required for binding to be GH-inducible (Hackett et al., 1995). A Box 1/Box 2 (GH-R aa 265–350) glutathione-S-transferase (GST) fusion protein additionally inhibited this GH-inducible DNA-protein interaction in a cell-free homogenate, but mutation of Y^{332} in this region had no effect, suggesting that tyrosine phosphorylation of the GH-R was not necessary for STAT activation in this system (Hackett et al., 1995). The precise biochemical mechanisms involved in GH-R regulated STAT transactivation thus remain to be elucidated. Figure 1 summarizes the structure-function studies described in this section.

ACTIVATION OF GENE EXPRESSION BY GROWTH HORMONE

The long-term effects of GH in animals are mediated by hormone-regulated changes in gene expression. To date, the transcription of several genes has been shown to be induced by GH, including c-*fos*; serine protease inhibitor genes, Spi 2.1 and 2.2; selected cytochrome p450 genes; and insulin-like growth factor-I (IGF-I).

Transcription of the early response gene, c-*fos*, is rapidly and transiently activated by GH and by several growth factors that regulate cellular mitogenesis and proliferation (Gurland et al., 1990; Ashcom et al., 1992). Regulation of c-*fos* transcription involves several *cis*-acting elements in the c-*fos* promoter, including the serum response element (SRE), the c-*sis* inducible element (SIE), an AP-1 site, an E-box, and a calcium/cAMP response element (Chen et al., 1995). At least two of these *cis*-acting elements play a major role in stimulating c-*fos* gene expression in response to GH. The SRE, which binds the serum response factor (SRF) and ternary complex factor (TCF), can mediate GH induction of c-*fos* gene expression in 3T3-F442A cells (Meyer et al., 1993). However, GH treatment does not regulate nuclear protein binding to the SRE (Gronowski et al., 1995). In contrast, a GH-inducible DNA-protein interaction occurs between the c-*fos* SIE and a protein immunologically related to STAT3 (Campbell et al., 1995; Gronowski et al., 1995). Transient transfection studies using a fusion gene containing the c-*fos* promoter and a measurable reporter demonstrated that both Box 1 of the GH-R and the NH_2-terminal domain of JAK2 must be intact for inducible promoter function (Frank et al., 1995); a functional JAK2 also was required. A chimeric GH-R lacking Box 1 fused to the JAK2 protein kinase domain could also stimulate c-*fos* promoter activity (Frank et al., 1995) (as noted earlier, the YGEFS motif in the extracellular portion of the GH-R is also necessary for GH-induced signaling, including activation of the c-*fos* promoter (Baumgartner et al., 1994)).

Two highly homologous rat hepatic serine protease inhibitor genes, Spi 2.1 and 2.2, are tightly controlled in a GH state-specific manner (Yoon et al., 1987). A closely related gene, Spi 2.3, escapes control by GH probably because of the absence of a GH-regulated *cis*-acting element in its proximal promoter, and perhaps also through a transcriptional repressor located in its unique 3' untranslated region (Le Cam and Legraverend, 1995). At least two independent GH-responsive *cis*-acting elements have been identified in the Spi 2.1 promoter, one that abuts the transcription start site (-41 to +8), and an upstream element, the GH response element (GHRE, -103 to -147). The proximal element confers a modest ~two-fold induction by GH, possibly through the presence of binding sites for the CCAAT enhancer-binding protein (C/EBP) transcription factor family (Le Cam et al., 1994). The GHRE is located within a GH-inducible DNase I hypersensitive site (Yoon et al., 1990). This region of DNA contains two GAS-like sequences that bind a protein termed the GH-inducible nuclear factor (GHINF) in a GH-dependent manner (Yoon et al., 1990; Sliva et al., 1994; Thomas et al., 1995).

GHINF activity is rapidly stimulated by GH administration *in vivo* (Thomas et al., 1995), is not blocked by the protein synthesis inhibitor cycloheximide (Gronowski et al., 1996), and is disrupted by phosphatases (Berry et al., 1994), suggesting that a reversible posttranslational modification mediates its ability to bind DNA. Recently, GHINF has been purified to near homogeneity, and has been shown to have an estimated mass of 92 kDa, similar to the size of STATs (Bergad et al., 1995). GHINF is recognized by a monoclonal antibody to the COOH-terminal region of STAT5, but it is unclear whether GHINF is an identical or related protein (Bergad et al., 1995). Co-transfection studies with a STAT5 expression vector stimulated activity of a fusion gene containing the Spi 2.1 GHRE, and mutation of the phosphorylated tyrosine (Y^{694}) in STAT5 blocked this induction (Wood et al., 1995). The regions of the GH-R that mediate transcriptional activation of Spi 2.1 include Box 1 and a segment of the cytoplasmic domain from aa 454 and 506 (Enberg et al., 1994; Goujon et al., 1994; Sotiropoulos et al., 1994; Billestrup et al., 1995), as indicated in Figure 1. The precise residues within the latter segment that are required for gene activation have not been identified.

GH also regulates the transcription of several members of the cytochrome p450 (CYP2C) family, including the rat liver-specific enzymes p450 2C7, 2C11, 2C12, and 2C13. At least two of these genes, CYP2C11 and CYP2C12, display sexually dimorphic patterns of gene expression in response to GH *in vivo* (Levgraverend et al., 1992; Sundseth et al., 1992). In male rats, the pulsatile secretion of GH promotes expression of the CYP2C11 gene, which encodes a steroid 16α- and 2α-hydroxylase. In the adult female rat, the higher frequency of GH secretion results in a relatively constant level of GH in the circulation, and positively regulates the female-specific CYP2C12 gene, a steroid disulfate 15β-hydroxylase. The *cis*-acting elements in these genes that confer differential responsiveness to GH are not known. However, recent studies on a related gene, cytochrome p450 3A10/lithochocholic acid 6 β-hydroxylase, have demonstrated the involvement of a STAT-like factor that bears partial immunoreactivity with STAT5 (Subramanian et al., 1995). Other recent evidence points to a role for cytosolic phospholipase A_2 in the regulation of 2CYP2C12 gene expression (Tollet et al., 1995). Phospholipase A_2, which is activated by ERKs through serine phosphorylation, generates the signaling molecule, arachidonic acid. This pathway is also augmented by calcium, and calcium inlux has been shown to be regulated by the ligand-activated GH-R (Billestrup et al., 1995).

GH administration to hypophysectomized rats rapidly stimulated IGF-I gene transcription through the coordinate use of its two promoters (Bichell et al., 1992; Le Stunff et al., 1995; Thomas et al., 1995); transcriptional activation was not blocked by the protein synthesis inhibitor, cycloheximide (Gronowski et al., 1996). Although the mechanism of transcriptional induction is unknown, a GH-inducible DNase I hypersensitive site has been mapped to the second intron of the IGF-I gene (Bichell et al., 1992; Thomas et al., 1995). Unlike the Spi 2.1 DNase I hypersensitive site, which binds a STAT-like factor in a GH-dependent manner, this region of the IGF-I gene appears to bind nuclear proteins constitutively (Thomas et al., 1995).

GH activates several additional transcription factors, potentially through secondary effects. Addition of GH to 3T3-F442A cells induced a rapid increase in the binding of activator protein (AP)-1, a heterodimer of c-*fos* and c-*jun*, to its cognate site in the metallothionein IIA gene promoter, and stimulated the binding of C/EBP to an oligonucleotide containing a high affinity C/EBP site (Clarkson et al., 1995). AP-1 activity is dependent upon ongoing c-*fos* protein synthesis, and was thus reduced by cycloheximide (Gronowski et al., 1996). GH modulates the expression of C/EBP isoforms by several mechanisms: GH stimulates translation of C/EBPβ mRNA (an effect dependent on tyrosine kinase, protein kinase A and protein kinase C-activated pathways), and induces C/EBPδ expression at the transcriptional level (an effect that is enhanced by cycloheximide (Clarkson et al., 1995)). The structure-function studies described in this section are also summarized in Figure 1.

GROWTH HORMONE RECEPTOR INTERNALIZATION AND TRANSLOCATION TO THE NUCLEUS

Although many cell surface hormone receptors are internalized, and then either degraded or recycled following ligand binding, there is increasing evidence that some cytokine receptors, including the GH-R, can be translocated to the nucleus, where they may exert additional, but as yet unknown, biological effects. The GH-R undergoes ligand-induced internalization. In the rat GH-R, internalization has been shown to be dependent on a domain between aa 318–380; mutation of F^{346} to A functionally disabled internalization, yet did not disrupt activation of a CAT reporter gene driven by the Spi 2.1 promoter (Allevato et al., 1995). Other aromatic amino acids between residues 318–380 could be mutated without effect, and deletion of Box 1 did not block GH-induced receptor internalization. These studies are summarized in Figure 1.

Similar regions of the GH-R appear to be required for translocation to the nucleus. The full length receptor has been found within the nucleus of many cell types *in vivo*, and has been shown to undergo rapid, ligand-dependent nuclear translocation (Lobie et al., 1994). In CHO cells transfected with the rat receptor, studies using ^{125}I-GH or an antibody directed against the intracellular domain of the GH-R demonstrated that the full length receptor (1–638) and a truncated mutant (1–454) were both associated with the nucleus; however, a GH-R truncated to amino acid 294 was not found in the nucleus, suggesting that residues 294–454 are necessary for nuclear translocation (Lobie et al., 1994). At present, no function has been described for nuclear GH-R.

OTHER SIGNAL TRANSDUCTION PATHWAYS ACTIVATED BY GROWTH HORMONE

The acute effects of GH on metabolism mimic the actions of insulin. Both hormones can stimulate glucose and amino acid uptake into cells and can inhibit lipolysis. In this regard, GH has been found to stimulate tyrosine phosphorylation of insulin receptor substrate-1 (IRS-1), a 185 kDa cytosolic protein that modulates many of the signal transduction pathways stimulated by insulin and IGF-I, including those involving intermediary metabolism. Phosphorylation of IRS-1 permits binding through different SH_2 domains of phosphotyrosine-containing proteins, including phosphatidylinositol (PI)-3 kinase, that modulate mitogenic signaling and stimulate the movement of glucose transporters to the cell surface. GH-inducible tyrosine phosphorylation of IRS-1 occurred in a time- and dose-dependent manner in primary adipocytes, and coincided with PI-3 kinase activation (Souza et al., 1994; Argetsinger et al., 1995; Ridderstråle et al., 1995). Activation of IRS-1 was lost with transfected truncated mutant GH receptors less than 380 aa in length, and deletion of Box 1 functionally impaired the ability of GH to stimulate IRS-1 or JAK2 tyrosine phosphoryaltion, suggesting that JAK2 may mediate tyrosine phosphorylation of IRS-1 (Argetsinger et al., 1995; Eriksson et al., 1995).

Other signal transducing molecules containing SH_2 domains such as SHC and GRB2, which have been postulated to link tyrosine kinase receptors to the MAP kinase pathway, also are activated by the GH-R. SHC is a group of cytoplasmic proteins consisting of at least three isoforms, 46 kDa, 52 kDa, and 66 kDa in size. SHC was originally discovered as a mediator of tyrosine kinase receptor signaling (Pelicci et al., 1992). SHC proteins are tyrosine phosphorylated after GH treatment, probably by JAK2, and deletion or mutation of Box 1 of the GH-R impaired the ability to phosphorylate SHC (VanderKuur et al., 1995). GRB2, a small cytosolic adaptor protein that can bind to SHC or IRS-1 through its SH_2 domains, can stimulate the RAS signaling pathway, and consequently MAP kinase activation, via an associated guanine nucleotide exchange protein, SOS. Although GRB2 does not bind to the GH-R or JAK2, its inducible binding to SHC or IRS-1 following GH treatment identifies a mechanism of GH-regulated MAP kinase activation (VanderKuur et al., 1995).

Calcium signaling is another pathway implicated in GH signal transduction, as noted earlier. GH stimulates opening of voltage-dependent L-type calcium channels in the plasma membranes of CHO cells stably transfected with the GH-R (Billestrup et al., 1995). Truncation of the GH-R proximal to aa 454 ablated calcium influx, but deletion or mutation of Box 1 did not affect this response, implying that JAK2 is not involved. A region of the GH-R between residues 454 and 506 is believed to contain the domain that triggers calcium influx, although the mechanism of activation is unknown (Billestrup et al., 1995). Although this same region of the GH-R has been linked to activation of Spi 2.1 transcription, there does not appear to be a direct connection between these two biological effects. GH-inducible calcium influx may lead to activation of phospholipase C.

SPECIFICITY OF SIGNALING BY THE GROWTH HORMONE RECEPTOR

One of the fundamental unsolved problems in GH research is identifying the specific signaling pathways used by the GH-R to regulate growth. Clearly, many cytokine receptors are capable of activating the same JAK, JAK2 which, in turn, can stimulate the phosphorylation of identical intracellular substrates, including STAT transcription factors. However, the biological outcomes differ dramatically among different cytokines. While some of the actions of GH are similar to those of prolactin, or to a lesser extent, to the effects of insulin, other actions, such as the activation of IGF-I gene expression and the stimulation of somatic growth, appear to be unique, suggesting that certain features of the GH-R must impart specificity. GH-R activation may involve the synergistic effects of several signal transduction pathways that converge at points downstream of the receptor, for example, at the level of target gene expression. Additionally, the temporal pattern of JAK- STAT activation, combined with stimulatory (or possibly inhibitory) effects on other signal transduction pathways, may define the more unique aspects of GH action. Finally, signal transduction pathways activated by GH *in vivo* may differ from effects observed *in vitro*. A current view of GH-regulated signaling is pictured in Figure 2. Although the results summarized in this figure indicate that major progress has been made in understanding the basics of GH-mediated signal transduction pathways, elucidation of the molecular mechanisms of GH action is still in its infancy.

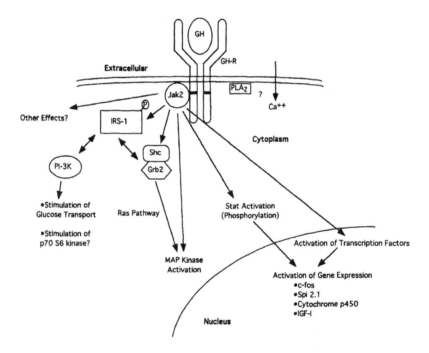

Figure 2. Summary of GH-R-mediated signal transduction. A single GH molecule binds to two GH-Rs, which dimerize and activate JAK2. The activation of JAK2 then initiates several intracellular events, including phosphorylation of intracellular proteins, STAT activation, MAP kinase activation, and phosphatidyl-inositol 3 kinase (PI-3K) activation. The mechanisms regulating stimulation of other transcription factors, leading to calcium influx, or activating phospholipase A_2 (PLA$_2$), are unknown.

SUMMARY

In the past five years, many advances have been made in our understanding of the molecular aspects of GH action. One GH molecule binds to two GH-Rs through two distinct sites on GH, and receptor dimerization is critical for signal transduction. A major subsequent step is the activation of JAK2, which leads to the phosphorylation and activation of several intracellular substrates, including MAP kinases and STATs. The induction of gene transcription by GH can occur through STATs or STAT-like factors, although alternative pathways of gene activation may be mediated by other transcription factors, such as AP-1 and C/EBP. Accessory signal transduction pathways that do not require JAK2 may modulate other biological effects of GH. These pleiomorphic pathways may synergize to produce specific actions of GH.

ACKNOWLEDGMENTS

The research described from the authors' laboratories is supported by NIH Research Grants R01-DK37449 and P01-HD20805.

REFERENCES

Allevato, G., Billestrup, N., Goujon, L., Galsgaard, E. D., Norstedt, G., Postel-Vinay, M.-C., Kelly, P. A., & Nielsen, J. H. (1995). Identification of phenylalanine 346 in the rat growth hormone receptor as being critical for ligand-mediated internalization and down-regulation. J. Biol. Chem. 270, 17210-17214.

Argetsinger, L. S., Campbell, G. S., Yang, X., Witthuhn, B. A., Silvennoinen, O., Ihle, J. N., & Carter-Su, C. (1993). Identification of JAK2 as a growth hormone receptor-associated tyrosine kinase. Cell 74, 237-244.

Argetsinger, L. S., Hsu, G. W., Meyers, M. E., Jr., Billestrup, N., White, M. F., & Carter-Su, C. (1995). Growth hormone, interferon-g, and leukemia inhibitory factor promoted tyrosyl phosphorylation of insulin receptor substrate-1. J. Biol. Chem. 270, 14685-14692.

Ashcom, G., Gurland, G., & Schwartz, J. (1992). Growth hormone synergizes with serum growth factors in inducing c-*fos* transcription in 3T3-F442A cells. Endocrinology 131, 1915-1921.

Baumbach, W. R., Horner, D. L., & Logan, J. S. (1989). The growth hormone-binding protein in rat serum is an alternatively spliced form of the rat growth hormone receptor. Genes Dev. 3, 1199-1205.

Baumgartner, J. W., Wells, C. A., Chen, C.-M., & Waters, M. J. (1994). The role of the WSXWS equivalent motif in growth hormone receptor function. J. Biol. Chem. 269, 29094-29101.

Bergad, P. L., Shih, H.-M., Towle, H. C., Schwarzenberg, S. J., & Berry, S. A. (1995). Growth hormone induction of hepatic serine protease inhibitor 2.1 transcription is mediated by a STAT5-related factor binding synergistically to two g-activated sites. J. Biol. Chem. 270, 24903-24910.

Berry, S. A., Bergad, P. L., Whaley, C. D., & Towle, H. C. (1994). Binding of a growth hormone-inducible nuclear factor is mediated by tyrosine phosphorylation. Mol. Endocrinol. 8, 1714-1719.

Bichell, D. P., Kikuchi, K., & Rotwein, P. (1992). Growth hormone rapidly activates insulin-like growth factor I gene transcription *in vivo*. Mol. Endocrinol. 6, 1899-1908.

Billestrup, N., Bouchelouche, P., Allevato, G., Ilondo, M., & Nielsen, J. H. (1995). Growth hormone receptor C-terminal domains required for growth hormone-induced intracellular free Ca^{2+} oscillations and gene transcription. Proc. Natl. Acad. Sci. USA 92, 2725-2729.

Burnside, J., & Cogburn, L. A. (1992). Developmental expression of hepatic growth hormone receptor and insulin-like growth factor-I mRNA in the chicken. Mol. Cell. Endocrinol. 89, 91-96.

Burnside, J., Liou, S. S., & Cogburn, L. A. (1991). Molecular cloning of the chicken growth hormone receptor complementary deoxyribonucleic acid: Mutation of the gene in sex-linked dwarf chickens. Endocrinology 128, 3183-3192.

Campbell, G. S., Christian, L. J., & Carter-Su, C. (1993). Evidence for involvement of the growth hormone receptor-associated tyrosine kinase in actions of growth hormone. J. Biol. Chem. 268, 7427-7434.

Campbell, G. S., Meyer, D. J., Raz, R., Levy, D. E., Schwartz, J., & Carter-Su, C. (1995). Activation of acute phase response factor (APRF)/STAT3 transcription factor by growth hormone. J. Biol. Chem. 270, 3974-3979.

Campbell, G. S., Pang, L., Miyasaka, T., Saltiel, A. R., & Carter-Su, C. (1992). Stimulation by growth hormone of MAP kinase activity in 3T3-F442A fibroblasts. J. Biol. Chem. 267, 6074-6080.

Chen, C., Clarkson, R. W. E., Xie, Y., Hume, D. A., & Waters, M. J. (1995a). Growth hormone and colony-stimulating factor 1 share multiple response elements in the c-*fos* promoter. Endocrinology 136, 4505-4516.

Chen, W. Y., Chen, N.-y., Yun, J., Wagner, T. E., & Kopchick, J. J. (1994). *In vitro* and *in vivo* studies of antagonistic effects of human growth hormone analogs. J. Biol. Chem. 269, 15892-15897.

Chen, W. Y., Chen, N.-y., Yun, J., Wight, D. C., Wang, X. Z., Wagner, T. E., & Kopchick, J. J. (1995a). Amino acid residues in the third a-helix of growth hormone involved in growth hormone promoting activity. Mol. Endocrin. 9, 292-302.

Clackson, T., & Wells, J. A. (1995). A hot spot of binding energy in a hormone-receptor interface. Science 267, 383-386.

Clarkson, R. W. E., Chen, C. M., Harrison, S., Wells, C., Muscat, G. E. O., & Waters, M. J. (1995). Early responses to *trans*-activating factors to growth hormone in preadipocytes: Differential regulation of CCAAT enhancer-binding protein-β (C/EBPβ) and C/EBPδ. Mol. Endocrinol. 9, 108-120.

Colosi, P., Wong, K., Leong, S. R., & Wood, W. I. (1993). Mutational analysis of the intracellular domain of the human growth hormone receptor. J. Biol. Chem. 268, 12617-12623.

Cunningham, B. C., Ultsch, M., Vos, A. M. d., Mulkerrin, M. G., Clauser, K. R., & Wells, J. A. (1991). Dimerization of the extracellular domain of the human growth hormone receptor by a single hormone molecule. Science 254, 821-825.

Darnell, J. E., Jr., Kerr, I. M., & Stark, G. R. (1994). Jak-STAT pathways and transcriptional activation in response to IFNs and other extracellular signaling proteins. Science 264, 1415-1421.

DaSilva, L., Howard, O. M. Z., Rui, H., Kirken, R. A., & Farrar, W. L. (1994). Growth signaling and JAK2 association mediated by membrane proximal cytoplasmic regions of prolactin receptors. J. Biol. Chem. 269, 18267-18270.

David, M., Petricoin, E., III, Benjamin, C., Pine, R., Weber, M. J., & Larner, A. C. (1995). Requirement for MAP kinase (ERK2) activity in interferon α- and interferon β-stimulated gene expression through STAT proteins. Science 269, 1721-1723.

de Vos, A. M., Ultsch, M., & Kossiakoff, A. A. (1992). Human growth hormone and extracellular domain of its receptor: crystal structure of the complex. Science 255, 306-312.

Dinerstein, H., Lago, F., Goujon, L., Ferrag, F., Esposito, N., Finidori, J., Kelly, P. A., & Postel-Vinay, M.-C. (1995). The proline-rich region of the GH receptor is essential for JAK2 phosphorylation, activation of cell proliferation, and gene transcription. Mol. Endocrinol. 9, 1701-1707.

Duquesnoy, P., Sobrier, M.-L., Duriez, B., Dastot, F., Buchanan, C. R., Savage, M. O., Preece, M. A., Craescu, C. T., Blouquit, Y., Goossens, M., & Anselem, S. (1994). A single amino acid substitution in the exoplasmic domain of the human growth hormone (GH) receptor confers familial GH resistance (Laron syndrome) with positive GH-binding activity by abolishing receptor homodimerization. EMBO J. 13, 1386-1395.

Enberg, B., Hulthén, A., Möller, C., Norstedt, G., & Francis, S. M. (1994). Growth hormone (GH) regulation of a rat serine protease inhibitor fusion gene in cells transfected with GH receptor cDNA. J. Mol. Endocrinol. 12, 39-46.

Eriksson, H., Ridderstråle, M., & Tornqvist, H. (1995). Tyrosine phosphorylation of the growth hormone (GH) receptor and janus tyrosine kinase-2 is involved in the insulin-like actions of GH in primary rat adipocytes. Endocrinology 136, 5093-5101.

Finbloom, D. S., Petricoin, E. F., III, Hackett, R. H., David, M., Feldman, G. M., Igarashi, K.-I., Fibach, E., Weber, M. J., Thorner, M. O., Silva, C. M., & Larner, A. C. (1994). Growth hormone and erythropoietin differentially activate DNA-binding proteins by tyrosine phosphorylation. Mol. Cell. Biol. 14, 2113-2118.

Frank, S. J., Gilliland, G., Kraft, A. S., & Arnold, C. S. (1994). Interaction of the growth hormone receptor cytoplasmic domain with the JAK2 tyrosine kinase. Endocrinology 135, 2228-2239.

Frank, S. J., Yi, W., Zhao, Y., Goldsmith, J. F., Gilliland, G., Jiang, J., Sakai, I., & Kraft, A. S. (1995). Regions of the JAK2 tyrosine kinase required for coupling to the growth hormone receptor. J. Biol. Chem. 270, 14776-14785.

Godowski, P. J., Leung, D. W., Meacham, L. R., Galgani, J. P., Hellmiss, R., Keret, R., Rotwein, P. S., Parks, J. S., Laron, Z., & Wood, W. I. (1989). Characterization of the human growth hormone receptor gene and demonstration of a partial gene deletion in two patients with Laron-type dwarfism. Proc. Nat. Acad. Sci. USA 86, 8083-8087.

Goujon, L., Allevato, G., Simonin, G., Paquereau, L., Cam, A. L., Clark, J., Nielsen, J. H., Dhane, J., Postel-Vinay, M.-C., Edery, M., & Kelly, P. A. (1994). Cytoplasmic sequences of the growth hormone receptor necessary for signal transduction. Proc. Natl. Acad. Sci. USA 91, 957-961.

Gronowski, A. M., & Rotwein, P. (1994). Rapid changes in nuclear protein tyrosine phosphorylation after growth hormone treatment *in vivo*: Identification of phosphorylated MAP kinase and STAT91. J. Biol. Chem. 269, 7874-7878.

Gronowski, A. M., Stunff, C. L., & Rotwein, P. (1996). Acute nuclear actions of growth hormone (GH): Cycloheximide inhibits inducible activator protein-1 activity, but does not block GH-regulated signal transducer and activator of transcription activation or gene expression. Endocrinology 137, 55-64.

Gronowski, A. M., Zhong, Z., Wen, Z., Thomas, M. J., Darnell, J. E., Jr., & Rotwein, P. (1995). *In vivo* growth hormone treatment rapidly stimulates the tyrosine phosphorylation and activation of Stat3. Mol. Endocrinology 9, 171-177.

Gurland, G., Ashcom, G., Cochran, B. H., & Schwartz, J. (1990). Rapid events in growth hormone action. Induction of c-*fos* and c-*jun* transcription in 3T3-F442A preadipocytes. Endocrinol. 127, 3187-3195.

Hackett, R. H., Wang, Y.-D., & Larner, A. C. (1995). Mapping of the cytoplasmic domain of the human growth hormone receptor required for activation of JAK2 and STAT proteins. J. Biol. Chem. 270, 21326-21330.

Horvath, C. M., Wen, Z., & Darnell, J. E., Jr. (1995). A STAT protein domain that determines DNA sequence recognition suggests a novel DNA-binding domain. Genes Dev. 9, 984-994.

Ihle, J. N. (1996). STATs: Signal transducers and activators of transcription. Cell 84, 331-334.

Ilondo, M. M., Damholt, A. B., Cunningham, B. A., Wells, J. A., Meyts, P. D., & Shymko, R. M. (1994). Receptor dimerization determines the effects of growth hormone in primary rat adipocytes and culture human IM-9 lymphocytes. Endocrinology 134, 2397-2403.

Kelly, P. A., Ali, S., Rozakis, M., Goujan, L., Nagano, M., Pellegrini, I., Gould, D., Djiane, J., Edery, M., Finidori, J., & Postel-Vinay, M. C. (1993). The growth hormone/prolactin receptor family. Recent Progr. Horm. Res. 48, 123-164.

Kishimoto, T., Taga, T., & Akira, S. (1994). Cytokine signal transduction. Cell 76, 253-256.

Kou, K., Lajara, R., & Rotwein, P. (1993). Amino acid substitutions in the intracellular part of the growth hormone receptor in a patient with the Laron syndrome. J. Clin. Endocrinol. Metab. 76, 54-59.

Laron, Z., Klinger, B., Eshet, R., Kaneti, H., Karasik, A., & Silbergeld, A. (1993). Laron syndrome due to a post-receptor defect: response to IGF-I treatment. Israel J. Med. Sci. 29, 757-763.

Le Cam, A., & Legraverend, C. (1995). Transcriptional repression, a novel function for 3' untranslated regions. Eur. J. Biochem. 231, 620-627.

Le Cam, A., Pantescu, V., Paquereau, L., Legraverend, C., Fauconnier, G., & Asins, G. (1994). *Cis*-acting elements controlling transcription from rat serine protease inhibitor 2.1 gene promoter. J. Biol. Chem. 269, 21532-21539.

Le Stunff, C., Thomas, M. J., & Rotwein, P. (1995). Rapid activation of rat insulin-like growth factor-I gene transcription by growth hormone reveals no changes in deoxyribonucleic acid-protein interactions within the second promoter. Endocrinology 136, 2230-2237.

Leung, D. W., Spencer, S. A., Cachianes, G., Hammonds, R. G., Collins, C., Henzel, W. J., Barnard, R., Waters, M. J., & Wood, W. I. (1987). Growth hormone receptor and serum binding protein: purification, cloning and expression. Nature 330, 537-543.

Levgraverend, C., Mode, A., Westin, S., Ström, A., Eguchi, H., Zaphiropoulos, P. G., & Gustafsson, J.-Å. (1992). Transcriptional regulation of rat P450 2C gene subfamily members by the sexually dimorphic pattern of growth hormone secretion. Mol. Endocrinology 6, 259-266.

Lobie, P. A., Allevato, G., Nielsen, J. H., Norstedt, G., & Billestrup, N. (1995). Requirement of tyrosine residues 333 and 338 of the growth hormone (GH) receptor for GH-stimulated function. J. Biol. Chem. 270, 21745-21750.

Lobie, P. E., Wood, T. J. J., Chen, C. M., Waters, M. J., & Norstedt, G. (1994). Nuclear translocation and anchorage of the growth hormone receptor. J. Biol. Chem. 269, 31735-31746.

Mathews, L. S., Enberg, B., & Norstedt, G. (1989). Regulation of rat growth hormone receptor gene expression. J. Biol. Chem. 264, 9905-9910.

Menon, R. K., Stephan, D. A., Singh, M., Morris, S. M., & Zou, L. (1995). Cloning of the promoter-regulatory region of the murine growth hormone receptor gene. J. Biol. Chem. 270, 8851-8859.

Meyer, D. J., Campbell, G. S., Cochran, B. H., Argetsinger, L. S., Larner, A. C., Finbloom, D. S., Carter-Su, C., & Schwartz, J. (1994). Growth hormone induces a DNA binding factor related to the interferon-stimulated 91-kD transcription factor. J. Biol. Chem. 269, 4701-4704.

Meyer, D. J., Stephenson, E. W., Johnson, L., Cochran, B. H., & Schwartz, J. (1993). The serum response element can mediate induction of c-*fos* by growth hormone. Proc. Natl. Acad. Sci. USA 90, 6721-6725.

Pelicci, G., Lanfrancone, L., Grignani, F., McGlade, J., Cavallo, F., Grignani, F., Pawson, T., & Pelicci, P. G. (1992). A novel transforming protein (SHC) with and SH2 domain is implicated in mitogenic signal transduction. Cell 70, 93-104.

Ridderstråle, M., Degerman, E., & Tornqvist, H. (1995). Growth hormone stimulates the tyrosine phosphorylation of the insulin receptor substrate-1 and its association with phosphatidylinositol 3-kinase in primary adipocytes. J. Biol. Chem. 270, 3471-3474.

Rosenfeld, R. G., Rosenbloom, A. L., & Guevara-Aguirre, J. (1994). Growth hormone (GH) insensitivity due to primary GH receptor deficiency. Endocrine Rev. 15, 369-390.

Rowlinson, S. W., Barnard, R., Bastiras, S., Robins, A. J., Brinkworth, R., & Waters, M. J. (1995). A growth hormone agonist produced by targeted mutagenesis at binding site 1: Evidence that site 1 regulates bioactivity. J. Biol. Chem. 270, 16833-16839.

Schindler, C., & Darnell, J. E., Jr. (1995). Transcriptional responses to polypeptide ligands: The Jak-STAT pathway. Annu. Rev. Biochem. 64, 621-651.

Silva, C. M., Lu, H., Weber, M. J., & Thorner, M. O. (1994a). Differential tyrosine phosphorylation of JAK1, JAK2, and STAT1 by growth hormone and interferon-g in IM-9 cells. J. Biol. Chem. 269, 27532-27539.

Sliva, D., Wood, T. J. J., Schindler, C., Lobie, P. E., & Norstedt, G. (1994b). Growth hormone specifically regulates serine protease inhibitor gene transcription via g-activated sequence-like DNA elements. J. Biol. Chem. 269, 26208-26214.

Smith, W. C., Linzer, D. H., & Talamantes, F. (1988). Detection of two growth hormone receptor mRNAs and primary translation products in the mouse. Proc. Natl. Acad. Sci. USA 85, 9576-9579.

Sotiropoulos, A., Perrot-Applanat, M., Dinerstein, H., Pallier, A., Postel-Vinay, M.-C., Finidori, J., & Kelly, P. A. (1994). Distinct cytoplasmic regions of the growth hormone receptor are required for activation of JAK2, mitogen-activated protein kinase, and transcription. Endocrinology 135, 1292-1298.

Souza, S. C., Frick, G. P., Yip, R., Lobo, R. B., Tai, L.-R., & Goodman, H. M. (1994). Growth hormone stimulates tyrosine phosphorylation of insulin receptor substrate-1. J. Biol. Chem. 269, 30085-30088.

Subramanian, A., Teixeira, J., Wang, J., & Gil, G. (1995). A STAT factor mediates the sexually dimorphic regulation of hepatic cytochrome P450 3A10/lithocholic acid beta hydroxylase gene expression by growth hormone. Mol. Cell. Biol. 15, 4672-4682.

Sundseth, S. S., Alberta, J. A., & Waxman, D. J. (1992). Sex-specific, growth hormone-regulated transcription of the cytochrome P450 2C11 and 2C12 gene. J. Biol. Chem. 267, 3907-3914.

Taniguchi, T. (1995). Cytokine signaling through nonreceptor protein tyrosine kinases. Science 268, 251-255.

Tanner, J. W., Chen, W., Young, R. L., Longmore, G. D., & Shaw, A. S. (1995). The conserved box 1 motif of cytokine receptors is required for association with JAK kinases. J. Biol. Chem. 270, 6523-6530.

Thomas, M. J., Gronowski, A. M., Berry, S. A., Bergad, P. L., & Rotwein, P. (1995). Growth hormone rapidly activates rat serine protease inhibitor 2.1 gene transcription and induces a DNA-binding activity distinct from those of Stat1, -3, and -4. Mol. Cell. Biol. 15, 12-18.

Thomas, M. J., Kikuchi, K., Bichell, D. P., & Rotwein, P. (1995a). Characterization of DNA-protein interactions at a growth hormone inducible nuclease hypersensitive site in the rat insulin-like growth factor-I gene. Endocrinology 136, 762-769.

Tollet, P., Hamberg, M., Gustafsson, J.-Å., & Mode, A. (1995b). Growth hormone signaling leading to CYP2C12 gene expression in rat hepatocytes involves phospholipase A_2. J. Biol. Chem. 270, 12569-12577.

Tourkine, N., Schindler, C., Larose, M., & Houdebine, L.-M. (1995). Activation of STAT factors by prolactin, interferon-g, growth hormone, and a tyrosine phosphatase inhibitor in rabbit primary mammary epithelial cells. J. Biol. Chem. 270, 20952-20961.

VanderKuur, J., Allevato, G., Billestrup, N., Norstedt, G., & Carter-Su, C. (1995a). Growth hormone-promoted tyrosyl phosphorylation of SHC proteins and SHC association with Grb2. J. Biol. Chem. 270, 7587-7593.

VanderKuur, J. A., Wang, X., Zhang, L., Allevato, G., Billestrup, N., & Carter-Su, C. (1995b). Growth hormone-dependent phosphorylation of tyrosine 333 and/or 338 of the growth hormone receptor. J. Biol. Chem. 270, 21738-21744.

VanderKuur, J. A., Wang, X., Zhang, Z., Campbell, G. S., Allevato, G., Billestrup, N., Norstedt, G., & Carter-Su, C. (1994). Domains of the growth hormone receptor required for association and activation of JAK2 tyrosine kinase. J. Biol. Chem. 269, 21709-21717.

Veals, S. A., Schindler, C., Leonard, D., Fu, X. Y., Aebersold, R., & Darnell, J. E., Jr. (1992). Subunit of an alpha-interferon-responsive transcription factor is related to interferon regulatory factor and Myb families of DNA-binding proteins. Mol. Cell. Biol. 12, 3315-3324.

Wang, X., Souza, S. C., Kelder, B., Cioffi, J. A., & Kopchick, J. J. (1995a). A 40-amino acid segment of the growth hormone receptor cytoplasmic domain is essential for GH-induced tyrosine-phosphorylated cytosolic proteins. J. Biol. Chem. 270, 6261-6266.

Wang, Y.-D., Wong, K., & Wood, W. I. (1995). Intracellular tyrosine residues of the human growth hormone receptor are not required for the signaling of proliferation or JAK-STAT activation. J. Biol. Chem. 270, 7021-7024.

Wang, Y.-D., & Wood, W. I. (1995b). Amino acid residues of the human growth hormone receptor that are required for proliferation and JAK-STAT signaling. Mol. Endocrinol. 9, 303-311.

Waxman, D. J., Ram, P. A., Park, S.-H., & Choi, H. K. (1995). Intermittent plasma growth hormone triggers tyrosine phosphorylation and nuclear translocation of a liver-expressed, STAT5-related DNA binding protein. J. Biol. Chem. 270, 13262-13270.

Wen, Z., Zhong, Z., & Darnell, J. E., Jr. (1995). Maximal activation of transcription by STAT1 and STAT3 requires both tyrosine and serine phosphorylation. Cell 82, 241-250.

Winston, L. A., & Bertics, P. J. (1992). Growth hormone stimulates the tyrosine phosphorylation of 42- and 45-kDa ERK-related proteins. J. Biol. Chem. 267, 4747-4751.

Wood, T. J. J., Sliva, D., Lobie, P. E., Pircher, T. J., Gouilleux, F., Wakao, H., Gustafsson, J.-Å., Groner, B., Norstedt, G., & Haldosén, L.-A. (1995). Mediation of growth hormone-dependent transcriptional activation by mammary gland factor/STAT5. J. Biol. Chem. 270, 9448-9453.

Yoon, J.-B., Berry, S. A., Seelig, S., & Towle, H. C. (1990). An inducible nuclear factor binds to a growth hormone-regulated gene. J. Biol. Chem. 265, 19947-19954.

Yoon, J.-B., Towle, H. C., & Seelig, S. (1987). Growth hormone induces two mRNA species of the serine protease inhibitor gene family in rat liver. J. Biol. Chem. 262, 4284-4289.

Zhang, X., Blenis, J., Li, H.-C., Schindler, C., & Chen-Kiang, S. (1995). Requirement of serine phosphorylation for formation of STAT-promoter complexes. Science 267, 1990-1994.

Chapter 3

Leptin

MARC REITMAN

Advances in Molecular and Cellular Endocrinology
Volume 2, pages 59-82.
Copyright © 1998 by JAI Press Inc.
All rights of reproduction in any form reserved.
ISBN: 0-7623-0292-5

INTRODUCTION

The discovery from Friedman's laboratory (Zhang et al., 1994) that a mutated leptin gene causes the massive obesity of the *ob/ob* mouse was a turning point in our understanding of the physiology of obesity and energy metabolism. This breakthrough was greeted with tremendous interest by the general public, exemplified by television reports and newspaper mention from the front to the comic pages. The business community was similarly enamored, with Amgen spending $20 million for the rights to the untested gene product. Given the level of excitement, it is perhaps understandable that scientists have swarmed to study this gene. This interest makes leptin (named from the Greek word for thin) a challenging but important topic to review.

Lep^{ob} is the official gene symbol for the mutated allele of the murine leptin gene, which was formerly named *ob* or *obese*. The human leptin gene symbol is *LEP*. *Lepr* and *LEPR* are the new symbols for the murine and (provisionally) human leptin receptor genes that were previously denoted *Obr* and *OB-R*, respectively. Thus the mutated *Lepr* allele formerly known as *db* or *diabetes* is now written $Lepr^{db}$. We will use the new nomenclature in this review.

THE LEP^{ob}/LEP^{ob} (ob/ob) MOUSE

Discovery of leptin followed from investigation of the Lep^{ob}/Lep^{ob} mouse (Ingalls et al., 1950; Coleman, 1978; Brey et al., 1990). These mice can reach 80 g (three times normal) with fat comprising 80% of body weight. Lep^{ob}/Lep^{ob} mice are unable to make functional leptin protein (see below). Their phenotype can be rationalized if the mutation causes the animals to act (metabolically and behaviorally) as if they are starving. The ensuing physiologic responses are inappropriate to the non-starving state and lead to early onset, massive obesity and diabetes.

In a field littered with discarded hypotheses for the primary defect (see Bray, 1996), the retrospectroscope confirms that experiments using parabiotic animals were providing important clues. Parabiotic mice are made by suturing the abdominal walls of two mice together, resulting in healthy mice with a small amount of crossed blood circulation. In Lep^{ob}/Lep^{ob} - +/+ parabiotic pairs, the +/+ mice showed no change in metabolism while the Lep^{ob}/Lep^{ob} mice ate less and had improved metabolic features. These data led to the suggestion that the +/+ mouse supplied a satiety/metabolic regulatory factor that was missing in the mutant mouse. Similar experiments led to the suggestion that the mutation in the *db/db* ($Lepr^{db}/Lepr^{db}$) mouse was in the gene encoding the receptor for this factor (Coleman, 1978).

Obesity in Lep^{ob}/Lep^{ob} mice results partially from increased food intake, a feature that is easy to quantitate and has received much attention. However, the obesity is also caused by decreased metabolic expenditure, since Lep^{ob}/Lep^{ob} mice gain more weight than pair-fed control animals (Coleman, 1978). Food restricted Lep^{ob}/Lep^{ob} mice deposit fat at the expense of lean body mass, suggesting that leptin has a role in energy partitioning (Dubuc, 1976). Lep^{ob}/Lep^{ob} mice have impaired sympathetic nervous system function, a lower body temperature (35 °C vs. 37 °C), lower thyroid hormone levels, and decreased physical activity, all of which contribute to reduced energy expenditure.

Lep^{ob}/Lep^{ob} mice are insulin resistant. Here the genetic background profoundly affects the phenotype. In C57BL/6J mice, the pancreatic islet β cells produce large amounts of insulin, resulting in islet hyperplasia, marked hyperinsulinemia, and milder diabetes. In contrast, C57BL/KsJ mice have transient hyperinsulinemia followed by β cell dysfunction, islet atrophy, less massive obesity, and severe diabetes.

Lep^{ob}/Lep^{ob} mice are hypogonadal and infertile with low follicle-stimulating hormone and luteinizing hormone production. It is likely that these mice have abnormal hypothalamic GnRH production rather than a primary pituitary defect. Transplantation of Lep^{ob}/Lep^{ob} ovaries into foster mothers allows fertility (Hummel, 1957), as does gonadotropin treatment of Lep^{ob}/Lep^{ob} females (Smithberg and Runner, 1957). Lep^{ob}/Lep^{ob} mice have elevated adrenocorticotropic hormone and glucocorticoid levels. Adrenalectomy improves glucose and insulin levels (Bailey et al., 1986) and prevents weight gain (Shimizu et al., 1993).

LEPTIN MOLECULAR BIOLOGY

Isolation of the leptin gene was achieved using positional cloning (Zhang et al., 1994). Briefly, the progeny of many matings were used to map the phenotype to a 650-kb region. This region was screened for genes having an adipose-specific expression pattern (the hypothesized expression pattern, which turned out to be correct). Proof that the isolated gene was the desired one required identification of the mutated base and characterization of its effect on protein levels.

Leptin's amino acid sequence is presented in Figure 1. The open reading frame predicts a preprotein of 167 amino acids and a secreted, mature protein of 146 amino acids (16.0 kDa). Five sequenced mammalian homologues are 85–91% identical to human leptin in the mature protein region. Although only mammalian leptins have been cloned, hybridization experiments suggest that other vertebrates have similar genes (Zhang et al., 1994). Leptin's sequence is not very similar to other known proteins, but an analysis of possible structural similarities suggests that it is a member of the helical cytokine family (Madej et al., 1995). Proteins in this family (growth hormone, interferons, and interleukins, among others) contain four α helices in a characteristic folding pattern. In leptin, the amino acids between putative helices 3 and 4 show the most variability between species. Aside from one disulfide bond, leptin does not undergo posttranslational modification (Cohen et al., 1996).

High levels of leptin mRNA are present in white and brown (Moinat et al., 1995) adipose tissue. The reports comparing leptin RNA levels in different fat deposits conflict, with no consensus for a consistent difference between locations (Masuzaki et al., 1995; Ogawa et al., 1995; Collins and Surwit, 1996; Harris et al., 1996). Cell lines able to differentiate into adipocytes (3T3-L1 and 3T3-F442A) do not express detectable leptin before differentiation and express low levels (up to a few percent of the level in mouse fat) after differentiation (MacDougald et al., 1995; Kallen and Lazar, 1996; Leroy et al., 1996; Rentsch and Chiesi, 1996). Leptin is also expressed at much lower levels in placenta and heart (Green et al., 1995). The importance of leptin production in nonadipose sites is just beginning to be examined.

The human leptin gene maps to chromosome 7q31.3 (Green et al., 1995) and to the syntenic regions of mouse chromosome 6 (10.5 cM) (Zhang et al., 1994). The human gene consists of three exons, a 29-bp first exon, a 171-bp second exon, and a 3.2-kb third exon with introns of 10.6 and 2.25 kb (Gong et al., 1996; Isse et al., 1995). The coding region is in the second and third exons. The murine genomic organization is similar to that in humans, and in the mouse two types of alternative RNA splicing have been observed: splicing three nucleotides into the third

```
                    20                    40        *           60                   80        90
                                    < helix 1 >                              < helix 2 >
human  MHWGTLCGFL WLWPYLFYVQ AVPIQKVQDD TKTLIKTIVT RINDISHTQS VSSKQKVTGL DFIPGLHPIL TLSKMDQTLA VYQQILTSMP
monkey       Y R  W              I          S                      R          V       Q        I    INL
mouse  C RP   R    S S                              A  R                      L       S
rat    C RP   R    S S            H                 AR R                      L       S
cow    RC P   R         E                              R                      V       S
sheep  ----------  ----------  ----------          S   R                      L       S  I    HA L
pig    RC P   R    S E     WR                   M      R                      V       S  I    L


                   100                   120                   140                 160       167
                          < helix 3 >                               < helix 4 >
human  SRNVIQISND LENL[R]DLLHV LAFSKSCHLP WASGLETLDS LGGVLEASGY STEVVALSRL QGSLQDMLWQ LDLSPGC
monkey                  L          E           D                              V E
mouse  Q L A      L          S    QT  QKP   D     L                          I Q
rat    Q L AH     L          S    QTR QKP   D     L                          I Q
cow               L          A    QVRA S     L     V
sheep             L          G    QVRA S E   V     E
pig    L ----     L          S    Q RA E     L     L ----------------- A     R
```

	% Identity to Human (aa 22–167)
monkey	91
mouse	87
rat	85
cow	90
sheep	--
pig	87

Figure 1. Amino acid sequence of mammalian leptins. Shown are the deduced sequences of human (U18915), Rhesus monkey (U58492), murine (U18812), rat (D45862), bovine (U43943), ovine (partial sequence, with missing data indicated by a dash; U62123), and swine (U59894) leptin. The complete human sequence is shown; for the rest only the differences are indicated. Arg 105 (reverse printed) is mutated to Thr in the Lep^{ob}/Lep^{ob} mouse. Due to alternate splicing, the codon for Gln 49 (*) is not present in some murine RNAs. The signal peptide is underlined. Regions predicted to be helical (Madej et al., 1995) are so noted. The percent identity in the mature protein (amino acids 22–167) to human leptin is also shown.

63

exon results in leptin missing glutamine 49 (the bioactivity of this protein has yet to be tested; it may be low) (Zhang et al., 1994). An extra noncoding exon is found between exons one and two in a small fraction of the mRNAs. Its inclusion appears to have no functional consequences (Mason et al., 1997).

The promoters of the human (Isse et al., 1995; Gong et al., 1996; Miller et al., 1996) and murine (He et al., 1995; de la Brousse et al., 1996; Hwang et al., 1996) leptin genes have been sequenced and studied by transient expression. In adipose cells, strong promoter activity required 109 bp (Mason et al., 1997), while only 84 bp was needed in HepG2 cells over-expressing C/EBPα (de la Brousse et al., 1996). The sequence includes a TATA box (at -30) and motifs that bind C/EBP (at -53), a novel factor (at -87), and Sp1 (at -96). The promoter is strongly transactivated by C/EBPα via the C/EBP site, consistent with this factor's importance for transcription of most adipocyte genes studied (de la Brousse et al., 1996; Hwang et al., 1996; Mason et al., 1997). In adipose cells the Sp1 site also contributes to promoter activity independent of the C/EBP site (Mason et al., 1997). No promoter binding sites for PPARγ have been identified and no distant enhancer elements have been reported. Similarly, no specific factors or promoter elements have been implicated in the regulation of leptin by adiposity or starvation.

LEPTIN RECEPTOR

The leptin receptor was cloned by screening an expression library for the ability to bind a tagged leptin molecule (Tartaglia et al., 1995). The receptor shows sequence similarity to the class I family of cytokine receptors (strongest to gp130 of the interleukin (IL)-6 receptor, the granulocyte colony stimulating factor receptor, and the leukemia inhibitory factor receptor). The extracellular portion has a ~200 amino acid region consisting of two fibronectin type III domains which probably constitute the ligand binding pocket. In the mouse, five different mRNA 3'-termini have been identified, arising via alternate splicing (Lee et al., 1996). The b isoform is the only one with an extensive cytoplasmic domain. Isoform b includes sequences (boxes 1, 2, and 3) that interact with the Janus kinase (JAK) family of tyrosine kinases. Three mRNAs (isoforms a, c, and d) encode shorter cytoplasmic tails, which are insufficient to interact with JAKs. The fifth isoform, e, encodes a protein truncated before the putative transmembrane region. In humans, four different leptin receptor 3' termini have been identified to date: the isoform b homologue, two species with short cytoplasmic tails (an isoform not described in mouse and an isoform a homologue), and an RNA that ends with a repetitive element (Cioffi et al., 1996). Two 5' ends for leptin receptor mRNAs have been reported (Tartaglia et al., 1995; Cioffi et al., 1996; Lee et al., 1996). Whether this is due to alternate promoter use and/or alternate RNA splicing remains to be determined, as do the details of the intron/exon organization. The total number of splice variants is not known, with more likely to be discovered. The leptin receptor gene maps to syntenic regions on human chromosome 1p31, mouse chromosome 4 (46.8 cM), and rat chromosome 5 (Truett et al., 1995; Chua et al., 1996a).

Leptin receptor RNA is widely expressed. In mouse, high levels are present in choroid plexus, lung, lymph node, and uterus, at moderate levels in kidney, and at lower levels in hypothalamus, heart, brain, spleen, liver, and muscle (Tartaglia et al., 1995; Cioffi et al., 1996; Ghilardi et al., 1996). Humans have detectable leptin receptor mRNA in most tissues, with higher levels in liver, heart, ovary, small intestine, and prostate (Cioffi et al., 1996). Binding studies confirmed that tissues expressing receptor RNA are also able to bind leptin (although such studies overestimate the number of receptors capable of signal transducer and activator of transcription (STAT) signal transduction).

Much can be predicted about leptin receptor signaling from its membership in the class I cytokine receptor family (Heldin, 1995; Darnell, 1996). These receptors are often activated by ligand- induced homo- or heterodimerization. This causes activation of receptor-associated JAKs, which phosphorylate tyrosines on STAT proteins. Phosphorylated STATs form homo- and heterodimers, translocate to the nucleus, and complex with other proteins. These STAT complexes bind to specific DNA motifs in promoters and enhancers through which they activate transcription.

Although leptin receptorology is in its infancy, some of the mechanistic details are being discovered. The receptor is a homodimer (Baumann et al., 1996) but the ligand:receptor stoichiometry is not known. Receptors with short cytoplasmic tails cannot signal (at least via the STAT pathway), and may interfere with STAT signaling by the full-length version (Baumann et al., 1996; Ghilardi et al., 1996). At this time the relevant JAKs have not been identified. Coexpression experiments suggest that the leptin receptor can activate STAT 1, 3, and 5 but not STAT6 (Baumann et al., 1996) or STAT 3, 5, and 6 but not STAT 1, 2, or 4 (Ghilardi et al., 1996). In more physiologic studies, *in vivo* leptin treatment caused activation of hypothalamic STAT3, but not STAT 1, 4, 5, or 6 in wild type and *Lep^{ob}/Lep^{ob}*, but not *Lep^{db}/Lep^{db}* mice (Vaisse et al., 1996). While some DNA motifs bound by STATs are known, other proteins in the complex may alter the binding specificity and, to date, no studies have identified specific target genes regulated by leptin stimulation.

Evidence that the long form of the leptin receptor performs a nonredundant function in the feed-back regulation of food intake and energy metabolism comes from demonstration that the *db* (*Lepr^{db}*) mutation is in this gene. The *Lepr^{db}* allele contains a point mutation that causes a premature stop codon in the isoform b RNA, due to inclusion of an extra 106 nucleotides (Chen et al., 1996; Lee et al., 1996). The *Lepr^{db-Pas}* mutation causes RNA levels of 5% of controls, due to partial duplication of the coding region (Chua et al., 1996a). The mutations in three other alleles (*Lepr^{db-2J}*, *Lepr^{db-3J}*, *Lepr^{db-ad}*) have not been identified. The rat mutation, *fa* (*fatty*) is also in the leptin receptor. One *fa* allele has a genomic rearrangement (Chua et al., 1996a). The other allele (*Lepr^{fa}*) has an A to C mutation, substituting Glu 269 by Pro. This mutation does not affect RNA levels, but causes decreased numbers of cell surface receptors (Chua et al., 1996b; Iida et al., 1996; Phillips et al., 1996).

What are the functions of the multiple carboxy-termini of the receptor? The JAK/STAT- functional receptor b isoform accounts for a minority of the receptor RNA, with relatively high levels of this isoform in the hypothalamus. The e isoform is predicted to encode a secreted protein. Does this protein function as a binding protein, similar to the insulin-like growth factor binding proteins? Do the receptors with short cytoplasmic regions transduce a signal via a non-JAK/STAT pathway, or inhibit JAK/STAT signals? Do the short forms transport leptin (such as across the blood/brain barrier) or increase leptin clearance, or do they buffer the tissue or plasma leptin levels (via binding thus decreasing clearance)? The evolutionary conservation and predominance of the short forms suggest that they will have important functions.

REGULATION OF LEPTIN IN CULTURED CELLS

The greatly increased leptin RNA in *Lep^{ob}/Lep^{ob}* fat suggests that RNA levels are regulated. Three types of mechanisms are likely to control leptin RNA levels. First, leptin production is tissue-specific, implicating tissue-selective transcription factors, such as C/EBPα (discussed above). Second, it is postulated that each cell's leptin RNA content reflects that cell's adiposity. For example, larger fat cells have more leptin RNA than smaller ones (Hamilton et al., 1995). This suggests that cell-autonomous mechanism(s) exist for sensing adiposity and converting the signal into leptin transcription. There is currently no experimental proof for this appealing idea. Third, hormonal and other metabolic regulators are likely to influence leptin RNA levels in a paracrine or endocrine manner.

Leptin RNA production has been measured in a number of cell culture systems but interpretation of the data has been hampered by the lack of an ideal model system and especially by the discordant responses seen in different model systems. While primary rat adipose cells maintain the adipose phenotype and remain viable in culture for more than one week, expression of some adipose-specific genes drops during the first day in culture (Gerrits et al., 1993). Leptin RNA levels decrease five-fold during this time, due to decreased transcription. The decrease is prevented by glucocorticoid treatment (Murakami et al., 1995; Slieker et al., 1996; He et al., 1997). Insulin increases leptin secretion and RNA levels, although both are small effects at early times (Saladin et al., 1995; Kolaczynski et al., 1996; Slieker et al., 1996; Barr et al., 1997). β-adrenergic agonists, particularly $β_3$ agonists, decrease leptin secretion and RNA levels. In addition, 17β-estradiol and adenosine agonists increase leptin protein levels while testosterone and aldosterone have no effect (Slieker et al., 1996).

3T3-L1 and 3T3-F442A cells can be induced to differentiate into adipocytes by hormone treatment. These lines express leptin RNA only after differentiation, and even then at levels ~100-fold below those in mouse adipose tissue (MacDougald et al., 1995; Kallen and Lazar, 1996; Leroy et al., 1996; Rentsch and Chiesi, 1996). Leptin RNA levels in differentiated 3T3-F442A cells drop 10-fold upon removal of the insulin required for differentiation and increase on insulin reintroduction (Leroy et al., 1996). In differentiated 3T3-L1 cells, insulin increased, thiazolidinediones and 2-bromopalmitate decreased leptin RNA levels, but dexamethasone, T_3, and isobutylmethylxanthine had no effect (Kallen and Lazar, 1996; Rentsch and Chiesi, 1996).

REGULATION OF LEPTIN IN HUMANS AND ANIMALS

Trying to integrate the available data on leptin regulation into a complete, coherent picture involves much fitting of square pegs into round holes.

While leptin administration to Lep^{ob}/Lep^{ob} mice demonstrates the hormone's satiety and energy expenditure effects, exactly how leptin fits into normal appetite and energy regulation is less clear. It is evident from many studies that the strongest predictor of blood leptin concentration and fat leptin RNA level is the total mass of body fat. Rodents with increased adiposity have high leptin, independent of the cause of the obesity (Frederich et al., 1995; Funahashi et al., 1995; Maffei et al., 1995). Specific examples include genetic mutations, hypothalamic lesions, diet, and treatment with toxic chemicals (gold thioglucose and monosodium glutamate). The only exception is, as expected, the Lep^{ob}/Lep^{ob} mouse, in which leptin is undetectable.

In humans, blood (serum or plasma) leptin concentrations range widely, from <0.5 to >100 ng/ml (<0.03 to >6.3 nM) (Maffei et al., 1995; Considine et al., 1996c; Ma et al., 1996; but see McGregor et al., 1996). Women have higher leptin levels than men with the same body mass index (BMI) (Maffei et al., 1995; Considine et al., 1996c). Some of the gender dimorphism is due to the larger amount of fat in women than men at a given BMI. However, gender-specific differences remain after correction for fat mass: plasma leptin is lowest in men, higher in postmenopausal women and highest in premenopausal women (Havel et al., 1996; Rosenbaum et al., 1996).

Serum leptin shows a two-fold diurnal variation. In humans, the highest levels are between midnight and 4 am, and lowest between noon and 4 pm (Sinha et al., 1996). Mice showed a similar periodicity (lowest at the beginning of the dark phase and highest at the end) without the inversion one might have expected in these nocturnal animals (Saladin et al., 1995; Ahima et al., 1996). Given the variability and the large range of blood leptin concentrations, it is perhaps surprising that large changes have not been observed with some experimental manipulations. For example, in humans there was no difference between leptin levels after an overnight fast compared with afternoon postprandial samples (Ma et al., 1996). Fasting causes a fall in murine and human leptin, but only to ~40% of the original level after 48 hours (Ahima et al., 1996; Boden et al., 1996). Since exogenous leptin has a half life of ~3 hours in the mouse, some leptin secretion probably continues during the fast (Ahima et al., 1996).

There are conflicting reports on the effects of glucose and insulin on leptin levels. Unclamped mice showed two-fold increases in leptin RNA 30 minutes after glucose or insulin treatment in seven-hour fasted animals (Mizuno et al., 1996). In humans, no change in leptin levels was seen with manipulation of blood glucose (oral or intravenous glucose tolerance or tolbutamide or insulin-induced hypoglycemia) after a 12-hour fast (Dagogo-Jack et al., 1996; Segal et al., 1996). No changes were seen in insulin-dependent diabetes mellitus or noninsulin-dependent diabetes mellitus (NIDDM) patients, or in euglycemic hyperinsulinemic clamp studies for up to five hours (Dagogo-Jack et al., 1996; Haffner et al., 1996; Kolaczynski et al., 1996; Vidal et al., 1996). A small increase (~two-fold) in serum leptin was observed by the end of a 72-hour hyperglycemic clamp (Kolaczynski et al., 1996). It seems plausible that the effects of glucose and insulin are relatively small in the fed state, but become more significant (either via direct or indirect mechanisms) in the starved state.

The β_3 adrenergic receptor is selectively present on adipocytes. Adipocyte β stimulation causes increased intracellular cAMP, leading to lipolysis in white and thermogenesis in brown adipose tissue. β_3 agonist treatment of mice causes a fall in leptin RNA and protein levels, notable after 12 hours of treatment (Mantzoros et al., 1996). Housing the animals at 4 °C also caused a drop in leptin levels, presumably via increased sympathetic activity to adipose tissue (Hardie et al., 1996).

Regulation of leptin by cytokines has been shown in the hamster and may contribute to infection-associated anorexia. Endotoxin increased leptin expression in fasted animals, presumably acting via tumor necrosis factor and IL-1 (Grunfeld et al., 1996). In general, there is a correlation between blood leptin levels and the amount of leptin RNA in adipose tissue. However, the plasma leptin seems to vary more than the adipose RNA levels. This is due, in part, to the fact that doubling of adipose tissue mass does not result in a doubling of blood volume. Some of the discrepancy may also be due to regulation of leptin by posttranscriptional mechanisms.

EFFECTS OF LEPTIN TREATMENT

On *Lep^{ob}/Lep^{ob}* Mice

Availability of recombinant leptin has allowed the effects of this hormone to be studied in mice. Large doses have been used (typically 1 to 10 mg/kg/day), possibly reflecting a lack of full bioactivity in the recombinant preparations. Unfortunately, use of large doses obscures the distinction between physiologic and pharmacologic effects.

Leptin treatment of Lep^{ob}/Lep^{ob} mice reverses every aspect of the *obese* phenotype examined (Campfield et al., 1995; Halaas et al., 1995; Pelleymounter et al., 1995; Stephens et al., 1995), including hyperphagia, weight gain, increased adiposity, decreased activity, decreased O_2 consumption, hypothermia, hyperinsulinemia and hyperglycemia, hypercorticosteronemia, increased hypothalamic neuropeptide Y (NPY) levels and NPY-induced hyperphagia (Smith et al., 1996), increased energy efficiency (Levin et al., 1996), decreased sympathetic stimulation of brown adipose tissue (Collins et al., 1996), and hypogonadism and infertility (Barash et al., 1996; Chehab et al., 1996). Leptin injected into the cerebral ventricles was more potent than hormone given intraperitoneally, suggesting a target with access to the cerebrospinal fluid (Campfield et al., 1995; Stephens et al., 1995). As expected, leptin treatment has no effect on $Lepr^{db}/Lepr^{db}$ mice that have a nonfunctional receptor (Campfield et al., 1995; Halaas et al., 1995; Stephens et al., 1995).

On Wild Type Mice

Leptin treatment of wild type mice produces less dramatic effects, but small reductions in food intake and body weight have been consistently observed (Campfield et al., 1995; Halaas et al., 1995; Pelleymounter et al., 1995). No adverse effects of leptin treatment have been reported. A crucial study from Flier's laboratory recently reported that leptin treatment of starved mice prevented some of the starvation-induced changes in thyroidal (reduced thyroxine), gonadal (reduced testosterone and luteinizing hormone; increased anestrous period), and adrenal function (increased corticosterone) (Ahima et al., 1996). This study stresses the importance of leptin in long (hours to days) term metabolic and hormonal regulation: a fall in blood leptin signals starvation.

On Isolated Adipose Cells

The effect of leptin on isolated cells has also been examined and is notable for the few effects seen. Treatment of rat adipocytes did not influence endogenous leptin secretion, suggesting the lack of an autocrine or paracrine feedback loop (Slieker et al., 1996). In adipocyte-like 30A5 cells, leptin appeared to decrease dexamethasone-induced acetyl-CoA carboxylase RNA and lipid synthesis (Bai et al., 1996). The paucity of experiments reporting effects of leptin on isolated cells may be due to the rarity of signaling-competent forms of the leptin receptor in the cells examined and to a bias against reporting negative results. With the cloning and characterization of the receptor, knowledge in this area should accrue rapidly.

Summary of Leptin's Effects

The current conventional wisdom model for leptin's action is that leptin is secreted from the adipose cell in proportion to its adiposity. This relationship is disrupted by starvation, during which leptin secretion drops. Thus blood leptin levels provide a signal available to the body about its degree of adiposity, as modified by its status regarding starvation. Leptin gets into the CSF, apparently via a saturable transport mechanism, with lower leptin levels found in the CSF than in blood (Caro et al., 1996; Schwartz et al., 1996). Probable leptin targets are receptors in the arcuate nucleus and hypothalamus, resulting in inhibition of NPY secretion (Stephens et al., 1995) (Figure 2). It seems likely that leptin also inhibits corticotropin releasing factor (CRF) secretion and stimulates gonadotropin releasing hormone (GnRH) and thyrotropin releasing hormone (TRH) secretion, since this would explain leptin's known effects on downstream hormones (Ahima et al., 1996). (Although it is also possible that leptin has a disproportionate effect on CRF with secondary effects on the gonadal and thyroidal axes.) Leptin also stimulates the sympathetic nervous system. Currently, the exact pathways and neurotransmitters for each of these effects remain to be elucidated. A good candidate is NPY, but mutant mice lacking NPY are neither thin nor leptin resistant (Erickson et al., 1996). While one can explain the phenotype of NPY knock out mice quite reasonably by proposing redundancy of the NPY-dependent pathways, the results are clearly not the predicted ones.

It is likely that physiologic functions of leptin remain to be discovered. The phenotype of the Lep^{ob}/Lep^{ob} mouse demonstrates that leptin is a non-redundant adiposity signal. However, other functions of leptin may not be obvious in Lep^{ob}/Lep^{ob} mice due to redundant pathways. In particular, the widespread distribution of leptin receptors suggests widespread actions. Perhaps we are being blinded by leptin's reputation as a satiety hormone.

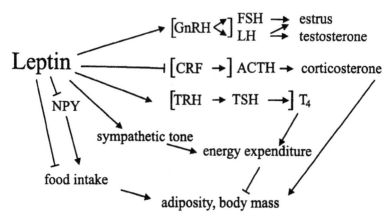

Figure 2. Effects of leptin. Stimulatory (→) and inhibitory (⊣) pathways are shown. Hypothetical effects are in brackets. Arrows may include multiple steps and intermediates. See text for further explanation. Abbreviations used: NPY, neuropeptide Y; GnRH, gonadotropin releasing hormone; FSH, follicle-stimulating hormone; LH, luteinizing hormone; CRF, corticotropin releasing factor; ACTH, adrenocorticotropic hormone; TRH, thyrotropin releasing hormone; and TSH, thyroid-stimulating hormone.

LEPTIN AND THE GENETICS OF HUMAN OBESITY

The Thrifty Genotype

It is well established that there is a large, polygenic hereditary contribution to human obesity (Stunkard et al., 1990). Yet obesity increases the risk of diabetes, hypertension, dyslipidemia, and weight-bearing osteoarthritis (Thomas, 1995). Why do humans have a genetic predisposition for this apparently deleterious trait? A plausible explanation is the thrifty genotype hypothesis: in times of scarcity those who could utilize a limited food supply more efficiently had a selective advantage, however in times of food affluence (such as our own) these genes are a liability (Neel, 1962). Indeed, the Lep^{ob} and Lep^{db} alleles confer resistance to starvation in both the homozygous and heterozygous states (Coleman, 1979).

Leptin Mutations

Do leptin mutations account for some human obesity? Linkage studies have shown a marginally significant linkage of the leptin region to severe obesity (BMI >35 in France or BMI >40 in the United States, correspond-ing to ~99th percentile), but not to milder obesity (Clement et al., 1996; Reed et al., 1996). Since genes other than leptin could account for the link-age, it will be important to elucidate the DNA sequences in this region that actually contribute to obesity. No human leptin mutations were found in screening the coding region in 105 obese people (Maffei et al., 1996) or in 149 obese and/or NIDDM subjects (Niki et al., 1996). One sequence varia-tion (Val94Met) without apparent functional significance has been re-ported (Considine et al., 1996b). Immunoreactive leptin was detectable in serum/plasma from 139 obese people (Considine et al., 1996c). These re-sults suggest that mutations in leptin account for, at most, a very small frac-tion of the human genetic predisposition to obesity.

The linkage study results have two possible explanations: this region either causes a weak predisposition in many people, or a strong predispo-sition in a few people. It is likely that, while very uncommon, human null mutations will eventually be found. By analogy to the murine Lep^{ob}/Lep^{ob} phenotype, we predict that the human clinical presentation will include severe obesity (e.g. BMI >50), comparatively lean parents and siblings (due to the autosomal recessive inheritance), onset of obesity early in life, a body temperature of 1–2 °C below normal, and decreased fertility. Pre-dicted laboratory findings include hypogonadotrophic hypogonadism, central hypothyroidism, and hypercortisolemia. Linkage and leptin se-quence analysis in a population of such individuals should prove particu-larly informative.

Leptin Receptor Mutations

Two lines of evidence suggest that the leptin receptor may be unusu-ally susceptible to mutation. First, receptor mutations arose multiple times and in two different species, for a total of seven independent events. Second, the numerous RNA splicing options may predispose to aberrant splicing when mutations occur in non-coding DNA (as apparently oc-curred in $Lepr^{db}$). Receptor mutations need not cause the extreme pheno-type such as seen in Lep^{ob}/Lep^{ob} mice, but could be quite mild (for example by decreasing the ratio of long to short isoforms). There are not yet published reports of mutations of the human leptin receptor. A number of silent nucleotide polymorphisms and an amino acid polymor-phism (Glu223Arg, which did not correlate with BMI) have been found (Considine et al., 1996a).

LEPTIN AND THE TREATMENT OF OBESITY

Leptin as a Drug

The abysmal failure of exhortations to diet, exercise, and change behavior to produce sustained weight loss has spurred a search for effective treatments for obesity. One possibility is long-term pharmacologic therapy. With the potential for a multi-billion dollar per year market, interest in leptin (and many other candidates) is high and recombinant human leptin is already in phase I human trials. What are the chances for success? Given the relatively rudimentary state of our understanding of the regulation of energy homeostasis, any answer is largely speculative.

Injection of recombinant leptin would be ideal therapy for patients with nonfunctional *leptin* genes, however such patients have yet to be discovered. While some obese people may have inappropriately low leptin levels for their amount of body fat, a more likely paradigm is that treatment will be with supra physiologic doses of leptin. An oft-made comparison is to insulin treatment of an already insulin-resistant NIDDM patient. It is likely that leptin treatment will not be a panacea and will not be without risk (although no obvious risks surfaced in the animal studies), but may prove useful especially when combined with diet, exercise, and possibly other pharmaceuticals (e.g., dexfenfluramine, phentermine, β_3 receptor agonists, serotorin 5HT2c receptor agonists, cholecystokinin-A receptor agonists, central glucagon-like peptide-1 agonists, NPY antagonists, urocortin agonists, drugs blocking fat uptake). Leptin treatment of human obesity is a reasonable, exciting experiment, the results of which are eagerly awaited.

Targeting the Leptin Pathway

What about anti-obesity drugs targeting the leptin-leptin receptor pathway? Agents that increase the magnitude of leptin secretion (as sulfonylureas do for insulin) would be desirable. Orally administrable leptin agonists are an obvious, but difficult, goal. (There still are no such insulin or cytokine agonists.) Pharmacologic agents acting immediately downstream of the leptin receptor are another theoretical possibility, but specificity may be a problem since the same JAKs and STATs are used for signaling by many other receptors.

CONCLUSIONS

The importance of leptin's discovery may be the insights it provides about the physiology and pathophysiology of the regulation of energy balance and adiposity, rather than any direct contribution as a genetic cause of or pharmacologic treatment for obesity. In the less than two years since the discovery of leptin a remarkable amount has been learned. However, there are some obvious gaps in our understanding.

- Leptin levels are regulated by adiposity, but we have no idea how the cell senses adiposity: is the level of a stored metabolite detected, is the energy status in the mitochondria sensed, is the physical size of the fat drop measured, or are there other mechanisms?
- Leptin levels drop upon starvation. What are the signals for this? Presumably this drop is not via a cell autonomous mechanism, so what hormones and/or metabolites are involved?
- What happens to leptin after it is secreted? Leptin is a sticky protein. Does it circulate bound to other proteins? Do such proteins affect its distribution to target organs, bioavailability, and clearance? What are the target sites outside the hypothalamus?
- Much about the leptin receptor remains to be defined. What genes are transcribed in response to leptin binding to its receptors? Which neural pathways are regulated by leptin to accomplish changes in energy balance? What are the functions of the receptors in tissues not relevant to energy balance? What are the functions of the majority of leptin receptors (the short forms) that cannot signal by the JAK/STAT pathways? Does genetic variation in the leptin receptor account for some predisposition to obesity?
- To what extent does leptin play a role in the normal control of reproductive endocrinology? Is leptin part of the signal that says the body has accumulated enough adipose stores to go through puberty?
- What about leptin physiology in species other than mammals? How widely distributed in evolution is leptin? Are there other hormones with similar functions to leptin, perhaps in other organisms? How widespread is leptin production in tissues besides fat, and what is the physiology?
- Obesity seems to be characterized by leptin resistance. Why does the obese individual not respond to high leptin levels with increased metabolic expenditure, decreased appetite, and weight loss? Are

there changes in intracellular leptin signal transduction, or is this a regulatory effect at the level of whole-body physiology?

There is much to learn.

ACKNOWLEDGMENTS

I am grateful to the members of my laboratory, for their contribution of ideas and support, and to Drs. V. Barr, O. Gavrilova, A. Ginsberg, M. Mason, and L. Weinstein for comments on the manuscript. This work was supported in part by a grant from the Lucille P. Markey Charitable Trust.

NOTE ADDED IN PROOF

Advances since the submission of this review include the crystal structure of leptin (Zhang et al., 1997) and the identification of humans with a nonfunctional leptin gene (Montague et al., 1997). Leptin is pulsatilely secreted (Licinio et al., 1997) and it may signal the onset of puberty (Mantzoros et al., 1997). In humans, leptin is elevated during pregnancy (Butte et al., 1997) due to placental production (Masuzaki et al., 1997) under the control of a tissue-specific enhancer (Bi et al., 1997).

Bi, S., Gavrilova, O., Gong, D.-W., Mason, M. M., and Reitman, M. (1997). Identification of a placental enhancer for the human leptin gene. J. Biol. Chem. 270, in press.
Butte, N. F., Hopkinson, J. M., and Nicolson, M. A. (1997). Leptin in human reproduction: serum leptin levels in pregnant and lactating women. J. Clin. Endo. Metab. 82, 585-589.
Licinio, J., Mantzoros, C., Negrao, A. B., Cizza, G., Wong, M. L., Bongiorno, P. B., Chrousos, G. P., Karp, B., Allen, C., Flier, J. S., and Gold, P. W. (1997). Human leptin levels are pulsatile and inversely related to pituitary-adrenal function. Nat. Med. 3, 575-579.
Mantzoros, C. S., Flier, J. S., and Rogol, A. D. (1997). A longitudinal assessment of hormonal and physical alterations during normal puberty in boys. V. Rising leptin levels may signal the onset of puberty. J. Clin. Endo. Metab. 82, 1066-1070.
Masuzaki, H., Ogawa, Y., Sagawa, N., Hosoda, K., Matsumoto, T., Mise, H., Nishimura, H., Yoshimasa, Y., Tanaka, I., Mori, T., and Nakao, K. (1997). Nonadipose tissue production of leptin: Leptin as a novel placenta-derived hormone in humans. Nat. Med. 3, 1029-1033.
Montague, C. T., Farooqi, I. S., Whitehead, J. P., Soos, M. A., Rau, H., Wareham, N. J., Sewter, C. P., Digby, J. E., Mohammed, S. N., Hurst, J. A., Cheetham, C. H., Earley, A. R., Barnett, A. H., Prins, J. B., and O'Rahilly, S. (1997). Congenital leptin

deficiency is associated with severe early-onset obesity in humans. Nature 387, 903-908.

Zhang, F., Basinski, M. B., Beals, J. M., Briggs, S. L., Churgay, L. M., Clawson, D. K., DiMarchi, R. D., Furman, T. C., Hale, J. E., Hsiung, H. M., Schoner, B. E., Smith, D. P., Zhang, X. Y., Wery, J. P., and Schevitz, R. W. (1997). Crystal structure of the obese protein leptin-E100. Nature 387, 206-209.

REFERENCES

Ahima, R.S., Prabakaran, D., Mantazoros, C., Qu, D., Lowell, B., Maratos-Flier, E., & Flier, J.S. (1996). Role of leptin in the neuroendocrine response to fasting. Nature 382, 250-252.

Bai, Y., Zhang, S., Kim, K.S., Lee, J.K., & Kim, K.H. (1996). Obese gene expression alters the ability of 30A5 preadipocytes to respond to lipogenic hormones. J. Biol. Chem. 271, 13939-13942.

Bailey, C.J., Day, C., Bray, G.A., Lipson, L.G., & Flatt, P.R. (1986). Role of adrenal glands in the development of abnormal glucose and insulin homeostasis in genetically obese (ob/ob) mice. Horm Metab Res. 18, 357-360.

Barash, I.A., Cheung, C.C., Weigle, D.S., Ren, H., Kabigting, E.B., Kuijper, J.L., Clifton, D.K., & Steiner, R.A. (1996). Leptin is a metabolic signal to the reproductive system. Endocrinology 137, 3144-3147.

Barr, V.A., Malide, D., Zarnowski, M.J., Cushman, S.W., & Taylor, S.I. (1997). Insulin stimulates both the secretion and the production of leptin by white fat tissue. Endocrinology, in press.

Baumann, H., Morella, K.K., White, D.W., Dembski, M., Bailon, P.S., Kim, H., Lai, C.F., & Tartaglia, L.A. (1996). The full-length leptin receptor has signaling capabilities of interleukin 6-type cytokine receptors. Proc. Natl. Acad. Sci. USA 93, 8374-8378.

Boden, G., Chen, X., Mozzoli, M and Ryan, I. (1996). Effect of fasting on serum leptin in normal human subjects. J. Clin. Endo. Metab. 81, 3419-3423.

Bray, G.A. (1996). Hereditary adiposity in mice: Human lessons from the yellow and obese (ob/ob) mice. Obesity Res. 4, 91-95.

Bray, G.A., Fisler, J., & York, D.A. (1990). Neuroendocrine control of the development of obesity: Understanding gained from studies of experimental animal models. Frontiers Neuroendocrinol. 11, 128-181.

Campfield, L.A., Smith, F.J., Guisez, Y., Devos, R., & Burn, P. (1995). Recombinant mouse OB protein: evidence for a peripheral signal linking adiposity and central neural networks. Science. 269, 546-549.

Caro, J.F., Kolaczynski, J.W., Nyce, M.R., Ohannesian, J.P., Opentanova, I., Goldman, W.H., Lynn, R.B., Zhang, P.L., Sinha, M.K., & Considine, R.V. (1996). Decreased cerebrospinal-fluid/serum leptin ratio in obesity: a possible mechanism for leptin resistance. Lancet 348, 159-161.

Chehab, F.F., Lim, M.E., & Lu, R. (1996). Correction of the sterility defect in homozygous obese female mice by treatment with the human recombinant leptin. Nat. Genet. 12, 318-320.

Chen, H., Charlat, O., Tartaglia, L.A., Woolf, E.A., Weng, X., Ellis, S.J., Lakey, N.D., Culpepper, J., Moore, K.J., Breitbart, R.E., Duyk, G.M., Tepper, R.I., & Morgenstern, J.P. (1996). Evidence that the diabetes gene encodes the leptin receptor: identification of a mutation in the leptin receptor gene in db/db mice. Cell 84, 491-495.

Chua, S.C., Jr., Chung, W.K., Wu-Peng, X.S., Zhang, Y., Liu, S.M., Tartaglia, L., & Leibel, R.L. (1996a). Phenotypes of mouse diabetes and rat fatty due to mutations in the OB (leptin) receptor. Science 271, 994-996.

Chua, S.C., Jr., White, D.W., Wu-Peng, X.S., Liu, S.M., Okada, N., Kershaw, E.E., Chung, W.K., Power-Kehoe, L., Chua, M., Tartaglia, L.A., & Leibel, R.L. (1996b). Phenotype of fatty due to Gln269Pro mutation in the leptin receptor (Lepr). Diabetes 45, 1141-1143.

Cioffi, J.A., Shafer, A.W., Zupancic, T.J., Smith-Gbur, J., Mikhail, A., Platika, D., & Snodgrass, H.R. (1996). Novel B219/OB receptor isoforms: possible role of leptin in hematopoiesis and reproduction. Nat. Med. 2, 585-589.

Clement, K., Garner, C., Hager, J., Philippi, A., LeDuc, C., Carey, A., Harris, T.J., Jury, C., Cardon, L.R., Basdevant, A., Demenais, F., Guy-Grand, B., North, M., & Froguel, P. (1996). Indication for linkage of the human OB gene region with extreme obesity. Diabetes 45, 687-690.

Cohen, S.L., Halaas, J.L., Friedman, J.M., Chait, B.T., Bennett, L., Chang, D., Hecht. R., & Collins, F. (1996). Human leptin characterization. Nature 382, 589.

Coleman, D.L. (1978). Obese and diabetes: two mutant genes causing diabetes-obesity syndromes in mice. Diabetologia 14, 141-148.

Coleman, D.L. (1979). Obesity genes: beneficial effects in heterozygous mice. Science 203, 663-665.

Collins, S., Kuhn, C.M., Petro, A.E., Swick, A.G., Chrunyk, B.A., & Surwit, R.S. (1996). Role of leptin in fat regulation. Nature 380, 677.

Collins, S., & Surwit, R.S. (1996). Pharmacologic manipulation of ob expression in a dietary model of obesity. J. Biol. Chem. 271, 9437-9440.

Considine, R.V., Considine, E.L., Williams, C.J., Hyde, T.M., & Caro, J.F. (1996a). The hypothalamic leptin receptor in humans: identification of incidental sequence polymorphisms and absence of the db/db mouse and fa/fa rat mutations. Diabetes 45, 992-994.

Considine, R.V., Considine, E.L., Williams, C.J., Nyce, M.R., Zhang, P., Opentanova, I., Ohannesian, J.P., Kolaczynski, J.W., Bauer, T.L., Moore, J.H., & Caro, J.F. (1996b). Mutation screening and identification of a sequence variation in the human ob gene coding region. Biochem. Biophys. Res. Commun. 220, 735-739.

Considine, R.V., Sinha, M.K., Heiman, M.L., Kriauciunas, A., Stephens, T.W., Nyce, M.R., Ohannesian, J.P., Marco, C.C., McKee, L.J., Bauer, T.L., & Caro, J.F. (1996c). Serum immunoreactive-leptin concentrations in normal-weight and obese humans. N. Engl. J. Med. 334, 292-295.

Dagogo-Jack, S., Fanelli, C., Paramore, D., Brothers, J., & Landt, M. (1996). Plasma leptin and insulin relationships in obese and nonobese humans. Diabetes 45, 695-698.

Darnell, J.E., Jr. (1996). Reflections on STAT3, STAT5, and STAT6 as fat STATs. Proc. Natl. Acad. Sci. USA 93, 6221-6224.

de la Brousse, F.C., Shan, B., & Chen, J.L. (1996). Identification of the promoter of the mouse obese gene. Proc. Natl. Acad. Sci. USA 93, 4096-4101.

Dubuc, P.U. (1976). Effects of limited food intake on the obese-hyperglycemic syndrome. Am. J. Physiol. 230, 1474-1479.

Erickson, J.C., Clegg, K.E., & Palmiter, R.D. (1996). Sensitivity to leptin and susceptibility to seizures of mice lacking neuropeptide Y. Nature 381, 415-421.

Frederich, R.C., Lollmann, B., Hamann, A., Napolitano-Rosen, A., Kahn, B.B., Lowell, B.B., & Flier, J.S. (1995). Expression of ob mRNA and its encoded protein in rodents. Impact of nutrition and obesity. J. Clin. Invest. 96, 1658-1663.

Funahashi, T., Shimomura, I., Hiraoka, H., Arai, T., Takahashi, M., Nakamura, T., Nozaki, S., Yamashita, S., Takemura, K., Tokunaga, K., & et al. (1995). Enhanced expression of rat obese (ob) gene in adipose tissues of ventromedial hypothalamus (VMH)-lesioned rats. Biochem. Biophys. Res. Commun. 211, 469-475.

Gerrits, P.M., Olson, A.L., & Pessin, J.E. (1993). Regulation of the GLUT4/muscle-fat glucose transporter mRNA in adipose tissue of insulin-deficient diabetic rats. J. Biol. Chem. 268, 640-644.

Ghilardi, N., Ziegler, S., Wiestner, A., Stoffel, R., Heim, M.H., & Skoda, R.C. (1996). Defective STAT signaling by the leptin receptor in diabetic mice. Proc. Natl. Acad. Sci. USA 93, 6231-6235.

Gong, D.W., Bi, S., Pratley, R.E., & Weintraub, B.D. (1996). Genomic structure and promoter analysis of the human obese gene. J. Biol. Chem. 271, 3971-3974.

Green, E.D., Maffei, M., Braden, V.V., Proenca, R., DeSilva, U., Zhang, Y., Chua, S.C., Leibel, R.L., Weissenbach, J., & Friedman , J.M. (1995). The human obese (OB) gene: RNA expression pattern and mapping on the physical, cytogenetic, and genetic maps of chromosome 7. Genome Res. 5, 5-12.

Grunfeld, C., Zhao, C., Fuller, J., Pollack, A., Moser, A., Friedman, J., & Feingold, K.R. (1996). Endotoxin and cytokines induce expression of leptin, the ob gene product, in hamsters. J Clin. Invest. 97, 2152-2157.

Haffner, S.M., Stern, M.P., Miettinen, H., Wei, M., & Gingerich, R.L. (1996). Leptin concentrations in diabetic and nondiabetic Mexican-Americans. Diabetes 45, 822-824.

Halaas, J.L., Gajiwala, K.S., Maffei, M., Cohen, S.L., Chait, B.T., Rabinowitz, D., Lallone, R.L., Burley, S.K., & Friedman, J.M. (1995). Weight-reducing effects of the plasma protein encoded by the obese gene. Science. 269, 543-546.

Hamilton, B.S., Paglia, D., Kwan, A.Y., & Deitel, M. (1995). Increased obese mRNA expression in omental fat cells from massively obese humans. Nat. Med. 1, 953-956.

Hardie, L.J., Rayner, D.V., Holmes, S., & Trayhurn, P. (1996). Circulating leptin levels are modulated by fasting, cold exposure and insulin administration in lean but not Zucker (fa/fa) rats as measured by ELISA. Biochem. Biophys. Res. Commun. 223, 660-665.

Harris, R.B., Ramsay, T.G., Smith, S.R., & Bruch, R.C. (1996). Early and late stimulation of ob mRNA expression in meal-fed and overfed rats. J. Clin. Invest. 97, 2020-2026.

Havel, P.J., Kasim-Karakas, S., Dubuc, G.R., Mueller, W., & Phinney, S.D. (1996). Gender differences in plasma leptin concentrations. Nat. Med. 2, 949-950.

He, Y., Mason, M.M., Chen, H., Quon, M.J., & Reitman, M. (1997). Unpublished observations.

He, Y., Chen, H., Quon, M.J., & Reitman, M. (1995). The mouse obese gene. Genomic organization, promoter activity, and activation by CCAAT/enhancer-binding protein alpha. J. Biol. Chem. 270, 28887-28891.

Heldin, C.H. (1995). Dimerization of cell surface receptors in signal transduction. Cell 80, 213-223.

Hummel, K.P. (1957). Transplantation of ovaries of the obese mouse. Anat. Rec. 128, 569.

Hwang, C.S., Mandrup, S., MacDougald, O.A., Geiman, D.E., & Lane, M.D. (1996). Transcriptional activation of the mouse obese (ob) gene by CCAAT/enhancer binding protein alpha. Proc. Natl. Acad. Sci. USA 93, 873-877.

Iida, M., Murakami, T., Ishida, K., Mizuno, A., Kuwajima, M., & Shima, K. (1996). Phenotype- linked amino acid alteration in leptin receptor cDNA from Zucker fatty (fa/fa) rat. Biochem. Biophys. Res. Commun. 222, 19-26.

Ingalls, A.M., Dickie, M.M., & Snell, G.D. (1950). Obese, a new mutation in the house mouse. J. Heredity 41, 317-318.

Isse, N., Ogawa, Y., Tamura, N., Masuzaki, H., Mori, K., Okazaki, T., Satoh, N., Shigemoto, M., Yoshimasa, Y., Nishi, S. et al. (1995). Structural organization and chromosomal assignment of the human obese gene. J. Biol. Chem. 270, 27728-27733.

Kallen, C.B., & Lazar, M.A. (1996). Antidiabetic thiazolidinediones inhibit leptin (ob) gene expression in 3T3-L1 adipocytes. Proc. Natl. Acad. Sci. USA 93, 5793-5796.

Kolaczynski, J.W., Nyce, M.R., Considine, R.V., Boden, G., Nolan, J.J., Henry, R., Mudaliar, S.R., Olefsky, J., & Caro, J.F. (1996). Acute and chronic effects of insulin on leptin production in humans: Studies in vivo and in vitro. Diabetes 45, 699-701.

Lee, G.H., Proenca, R., Montez, J.M., Carroll, K.M., Darvishzadeh, J.G., Lee, J.I., & Friedman, J.M. (1996). Abnormal splicing of the leptin receptor in diabetic mice. Nature 379, 632-635.

Leroy, P., Dessolin, S., Villageois, P., Moon, B.C., Friedman, J.M., Ailhaud, G., & Dani, C. (1996). Expression of ob gene in adipose cells. Regulation by insulin. J. Biol. Chem. 271, 2365-2368.

Levin, N., Nelson, C., Gurney, A., Vandlen, R., & de Sauvage, F. (1996). Decreased food intake does not completely account for adiposity reduction after ob protein infusion. Proc. Natl. Acad. Sci. USA 93, 1726-1730.

Ma, Z., Gingerich, R.L., Santiago, J.V., Klein, S., Smith, C.H., & Landt, M. (1996). Radioimmunoassay of leptin in human plasma. Clin. Chem. 42, 942-946.

MacDougald, O.A., Hwang, C.S., Fan, H., & Lane, M.D. (1995). Regulated expression of the obese gene product (leptin) in white adipose tissue and 3T3-L1 adipocytes. Proc. Natl. Acad. Sci. USA 92, 9034-9037.

Madej, T., Boguski, M.S., & Bryant, S.H. (1995). Threading analysis suggests that the obese gene product may be a helical cytokine. FEBS Lett. 373, 13-18.

Maffei, M., Halaas, J., Ravussin, E., Pratley, R.E., Lee, G.H., Zhang, Y., Fei, H., Kim, S., Lallone, R., Ranganathan, S. et al. (1995). Leptin levels in human and rodent: measurement of plasma leptin and ob RNA in obese and weight-reduced subjects. Nat. Med. 1, 1155-1161.

Maffei, M., Stoffel, M., Barone, M., Moon, B., Dammerman, M., Ravussin, E., Bogardus, C., Ludwig, D.S., Flier, J.S., Talley, M., Auerbach, S., & Friedman, J.M. (1996). Absence of mutations in the human OB gene in obese/diabetic subjects. Diabetes 45, 679-682.

Mantzoros, C.S., Qu, D., Frederich, R.C., Susulic, V.S., Lowell, B.B., Maratos-Flier, E., & Flier, J.S. (1996). Activation of beta(3) adrenergic receptors suppresses leptin expression and mediates a leptin-independent inhibition of food intake in mice. Diabetes. 45, 909-914.

Mason, M.M., He, Y., Chen, H., Quon, M.J., & Reitman, M. (1997). Regulation of leptin promoter function by Spl, C/EBP, and a novel factor. Manuscript submitted..

Masuzaki, H., Ogawa, Y., Isse, N., Satoh, N., Okazaki, T., Shigemoto, M., Mori, K., Tamura, N., Hosoda, K., Yoshimasa, Y. et al. (1995). Human obese gene expression. Adipocyte- specific expression and regional differences in the adipose tissue. Diabetes 44, 855-858.

McGregor, G.P., Desaga, J.F., Ehlenz, K., Fischer, A., Heese, F., Hegele, A., Lammer, C., Peiser, C., & Lang, R.E. (1996). Radiommunological measurement of leptin in plasma of obese and diabetic human subjects. Endocrinology 137, 1501-1504.

Miller, S.G., De Vos, P., Guerre-Millo, M., Wong, K., Hermann, T., Staels, B., Briggs, M.R., & Auwerx, J. (1996). The adipocyte specific transcription factor C/EBPalpha modulates human ob gene expression. Proc. Natl. Acad. Sci. USA 93, 5507-5511.

Mizuno, T.M., Bergen, H., Funabashi, T., Kleopoulos, S.P., Zhong, Y.G., Bauman, W.A., & Mobbs, C.V. (1996). Obese gene expression: reduction by fasting and stimulation by insulin and glucose in lean mice, and persistent elevation in acquired (diet-induced) and genetic (yellow agouti) obesity. Proc. Natl. Acad. Sci. USA 93, 3434-3438.

Moinat, M., Deng, C., Muzzin, P., Assimacopoulos-Jeannet, F., Seydoux, J., Dulloo, A.G., & Giacobino, J.P. (1995). Modulation of obese gene expression in rat brown and white adipose tissues. FEBS Lett. 373, 131-134.

Murakami, T., Iida, M., & Shima, K. (1995). Dexamethasone regulates obese expression in isolated rat adipocytes. Biochem. Biophys. Res. Commun. 214, 1260-1267.

Neel, J.V. (1962). Diabetes mellitus: a "thrifty" genotype rendered detrimental by "progress"? Am. J. Hum. Genet. 14, 353-362.

Niki, T., Mori, H., Tamori, Y., Kishimoto-Hashimoto, M., Ueno, H., Araki, S., Masugi, J., Sawant, N., Majithia, H.R., Rais, N., Hashiramoto, M., Taniguchi, H., & Kasuga, M. (1996). Human obese gene: molecular screening in Japanese and Asian Indian NIDDM patients associated with obesity. Diabetes 45, 675-678.

Ogawa, Y., Masuzaki, H., Isse, N., Okazaki, T., Mori, K., Shigemoto, M., Satoh, N., Tamura, N., Hosoda, K., Yoshimasa, Y. et al. (1995). Molecular cloning of rat obese cDNA and augmented gene expression in genetically obese Zucker fatty (fa/fa) rats. J. Clin. Invest. 96, 1647-1652.

Pelleymounter, M.A., Cullen, M.J., Baker, M.B., Hecht, R., Winters, D., Boone, T., & Collins, F. (1995). Effects of the obese gene product on body weight regulation in ob/ob mice. Science 269, 540-543.

Phillips, M.S., Liu, Q., Hammond, H.A., Dugan, V., Hey, P.J., Caskey, C.J., & Hess, J.F. (1996). Leptin receptor missense mutation in the fatty Zucker rat. Nat. Genet. 13, 18-19.

Reed, D.R., Ding, Y., Xu, W., Cather, C., Green, E.D., & Price, R.A. (1996). Extreme obesity may be linked to markers flanking the human OB gene. Diabetes 45, 691-694.

Rentsch, J., & Chiesi, M. (1996). Regulation of ob gene mRNA levels in cultured adipocytes. FEBS Lett. 379, 55-59.

Rosenbaum, J., Nicolson, M., Hirsch, J., Heymsfield, S.B., Gallagher, D., Chu, F., & Leibel, R.L. (1996). Effects of gender, body composition, and menopause on plasma concentrations of leptin. J. Clin. Endo. Metab. 81, 3424-3427.

Saladin, R., De Vos, P., Guerre-Millo, M., Leturque, A., Girard, J., Staels, B., & Auwerx, J. (1995). Transient increase in obese gene expression after food intake or insulin administration. Nature 377, 527-529.

Schwartz, M.W., Peskind, E., Raskind, M., Boyko, E.J., & Porte, D., Jr. (1996). Cerebrospinal fluid leptin levels: relationship to plasma levels and to adiposity in humans. Nat. Med. 2, 589-593.

Segal, K.R., Landt, M., & Klein, S. (1996). Relationship between insulin sensitivity and plasma leptin concentration in lean and obese men. Diabetes. 45, 988-991.

Shimizu, H., Ohishima, K., Bray, G.A., Peterson, M., & Swerdloff, R.S. (1993). Adrenalectomy and castration in the genetically obese (ob/ob) mouse. Obesity Res. 1, 377-383.

Sinha, M.K., Ohannesian, J.P., Heiman, M.L., Kriauciunas, A., Stephens, T.W., Magosin, S., Marco, C., & Caro, J.F. (1996). Nocturnal rise of leptin in lean, obese, and non-insulin- dependent diabetes mellitus subjects. J. Clin. Invest. 97, 1344-2347.

Slieker, L.J., Sloop, K.W., Surface, P.L., Kriauciunas, A., LaQuier, F., Manetta, J., Bue-Valleskey, J., & Stephens, T.W. (1996). Regulation of expression of ob mRNA and protein by glucocorticoids and cAMP. J. Biol. Chem. 271, 5301-5304.

Smith, F.J., Campfield, L.A., Moschera, J.A., Bailon, P.S., & Burn, P. (1996). Feeding inhibition by neuropeptide Y. Nature 382, 307.

Smithberg, M., & Runner, M.N. (1957). Preganancy induced in genetically sterile mice. J Heredity 48, 97-100.

Stephens, T.W., Basinski, M., Bristow, P.K., Bue-Valleskey, J.M., Burgett, S.G., Craft, L., Hale, J., Hoffmann, J., Hsiung, H.M., Kriauciunas, A. et al. (1995). The role of neuropeptide Y in the antiobesity action of the obese gene product. Nature 377, 530-532.

Stunkard, A.J., Harris, J.R., Pedersen, N.L., & McClearn, G.E. (1990). The body-mass index of twins who have been reared apart. N. Engl. J. Med. 322, 1483-1487.

Tartaglia, L.A., Dembski, M., Weng, X., Deng, N., Culpepper, J., Devos, R., Richards, G.J., Campfield, L.A., Clark, F.T., Deeds, J. et al. (1995). Identification and expression cloning of a leptin receptor, OB-R. Cell 83, 1263-1271.

Thomas, P.R. (1995). Weighing the Options. National Academy Press, Washington, DC.

Truett, G.E., Jacob, H.J., Miller, J., Drouin, G., Bahary, N., Smoller, J.W., Lander, E.S., & Leibel, R.L. (1995). Genetic map of rat chromosome 5 including the fatty (fa) locus. Mamm. Genome. 6, 25-30.

Vidal, H., Auboeuf, D., De Vos, P., Staels, B., Riou, J.P., Auwerx, J., & Laville, M. (1996). The expression of ob gene is not acutely regulated by insulin and fasting in human abdominal subcutaneous adipose tissue. J. Clin. Invest. 98, 251-255.

Vaisse, C., Halaas, J.L., Horvath, C.M., Darnell, J.E., Jr., Stoffel, M., & Friedman, J.M. (1996). Leptin activation of Stat3 in the hypothalamus of wild-type and ob/ob mice but not db/db mice. Nat. Genet. 14, 95-97.

Zhang, Y., Proenca, R., Maffei, M., Barone, M., Leopold, L., & Friedman, J.M. (1994). Positional cloning of the mouse obese gene and its human homologue. Nature 372, 425-432.

Chapter 4

Cellular Mechanisms of Signal Transduction for Growth Factors

ALAN R. SALTIEL

Advances in Molecular and Cellular Endocrinology
Volume 2, pages 83-97.
Copyright © 1998 by JAI Press Inc.
All rights of reproduction in any form reserved.
ISBN: 0-7623-0292-5

INTRODUCTION

Growth factors exert profound effects on the metabolism, survival, and differentiation of numerous cell types. Studies on the genetics and physiology of development have revealed that the synthesis of these factors is under tight control, and furthermore that their functions are highly specialized, yet the receptors for these molecules appear to be surprisingly similar in structure. Growth factor receptors have three distinct domains, an extracellular ligand binding region, a hydrophobic transmembrane region, and a cytoplasmic region that often contains tyrosine kinase activity. These receptor tyrosine kinases (RTK), which comprise a gene superfamily of over 30 members, are similarly activated upon ligand binding, undergoing dimerization and autophosphorylation on tyrosine residues that initiates a series of distinct, but often redundant, signaling pathways. Growth factors induce a variety of changes in transport properties; acute metabolic activities; membrane trafficking; cytoskeletal interactions; and a program of gene induction governing cell growth, differentiation, and migration. Some of the factors induce a program of cell growth and viability in some cells and differentiation in others. Interestingly, growth factors with quite different cellular effects induce many of the same signaling pathways upon receptor binding, suggesting that subtle differences result from combinations of divergent and convergent signaling pathways, invoking complex mechanisms that allow cells to respond appropriately.

This review will highlight some of the current advances in signal transduction initiated by growth factor tyrosine kinase receptors. This is not intended as an exhaustive or comprehensive survey of receptors or their potential signaling pathways, and will not include discussions of cross regulation that may occur between tyrosine kinases and other well established effector systems governing cell growth involving ion channels, cytoskeletal elements, phospholipids or other protein kinases and phosphatases. Rather, it will focus on the molecular interactions involved in receptor activation, the early events that produce specificity in signal transduction, and how these signals are translated into downstream regulation of serine kinase cascades as an example of one pathway that ultimately regulates gene expression to control cellular growth and differentiation.

MOLECULAR INTERACTIONS OF GROWTH FACTOR RECEPTORS

Receptors Are Activated by Tyrosine Phosphorylation

The gaps in our understanding of signal transduction for RTKs have been narrowed by recent insights into protein-protein interactions produced as a result of receptor activation. It is generally believed that most of these receptors undergo a similar program of dimerization and transactivation following ligand binding, resulting in tyrosine phosphorylation (Figure 1) (Koch et al., 1991; Schlessinger and Ullrich, 1992). In turn, these tyrosine phosphorylated proteins serve as binding sites for a number of functionally related signaling molecules. Such interactions are typified by those with proteins containing *src* homology (SH) 2 domains (Koch et al., 1991). These are conserved noncatalytic regions of approximately 100 amino acids that function as phosphotyrosine recognition motifs for specific sequences. Some of these proteins, like the

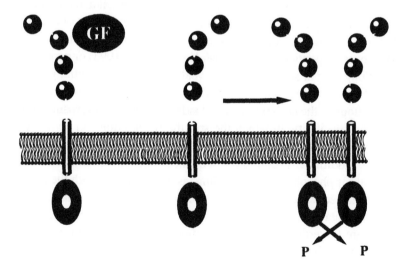

Figure 1. Growth factor receptors undergo a program of dimerization and autophosphorylation upon binding. After growth factors bind to their receptors on the cell surface, they are thought to form homodimers. Dimerization results in the activation of the tyrosine kinase, leading to the intermolecular phosphorylation on tyrosine residues in the carboxy-terminal tails of the receptor.

pp60^{src} tyrosine kinase itself, as well as phospholipase C (PLC)γ, *ras* GTPase activating protein (GAP), the protein tyrosine phosphatase (PTPase) SHPTP2 or Syp, and the oncoprotein *vav*, have catalytic activities that are involved in signaling pathways. These proteins can be tyrosine phosphorylated upon binding to RTKs in some cases, although the functional consequences are unclear. For other SH2 proteins, including the 85 kd regulatory subunit of phosphatidylinositol-(PI)-3 kinase, Grb-2, and the oncoproteins *crk* and *nck*, catalytic activities have not been detected, suggesting that these molecules serve as adapter proteins, linking receptors or their substrates to other pathways. In all cases, the binding of SH2 proteins is absolutely dependent upon tyrosine phosphorylation of the receptor, and is mediated solely through the SH2 domain (Downing et al., 1989; Anderson et al., 1990; Reedijk et al., 1990; Escobedo et al., 1991; Hu et al., 1992; Ohmichi et al., 1992a).

The structural basis for differential association of RTKs with SH2 domains is now fairly well understood (Kuriyan and Cowburn, 1993). Direct binding measurements have revealed quantitative kinetic differences in these interactions, perhaps accounting for differential activation of pathways that are dictated by receptor concentration. Analyses of competition binding among SH2 domains to mutated receptors (Zhu et al., 1992), or studies using peptide inhibitors (McNamara et al., 1993) or peptide libraries (Songyang et al., 1993) have revealed certain preferential primary sequence motifs in receptors for SH2 binding. In general, the amino acids contiguous to the carboxy-terminal side of the phosphotyrosine prescribe binding specificity, in which the specific recognition can be dictated by residues 2–5 positions downstream from the tyrosine (Songyang et al., 1993). Additionally, conserved sequences in SH2 domains appear to be critical for phosphotyrosine binding, as demonstrated with the conserved FLVRES sequence (Mayer et al., 1992; Zhu et al., 1993). Analysis of these binding data for the *abl* SH2 domain, considered in light of the solution structure (Overduin et al., 1992), suggested that the two primary amino groups from the conserved *Arg* residue undergo bidentate interactions with two of the oxygen atoms of the phosphate, while surrounding *Ser* hydroxyls may hydrogen bind the remaining oxygen.

There are likely to be other protein interaction domains found in proteins that interact with RTKs. One such domain was identified recently in the tyrosine kinase substrates Shc and IRS-1, known as the PTB or PI domain (Kavanaugh et al., 1995; van der Geer and Pawson, 1995). These domains are functionally similar to SH2 domains, in that they recognize sequences containing phosphotyrosine, but appear to have a different sequence specificity, in which residues amino terminal to the tyrosine are important for binding. Many of the proteins that interact with Shc through this domain have the sequence NPXY(p) (Kavanaugh et al., 1995; van der Geer and Pawson, 1995), although sequence variations further upstream are likely to further influence binding affinity. Recent structural studies (Zhou et al., 1995) suggest that these domains may resemble PH domains, another motif commonly found in signaling proteins.

Receptor: SH2 Interactions Generate a Variety of Signaling Pathways

Although we have a reasonably in depth understanding of the three-dimensional structure and molecular dynamics for many examples of RTK-SH2 or PTB binding, the precise functions of these interactions have been more difficult to resolve. In some cases, SH2 domain occupancy directly modulates the activity of the protein, as is the case for the tyrosine phosphatase SHPTP2 or Syp, in which catalytic activity is markedly elevated by the interaction of the phosphatase with tyrosine phosphorylated receptors or their substrates (Milarski and Saltiel, 1994; Ploskey et al., 1995). In other cases, SH2-and PTB-containing proteins are tyrosine phosphorylated when bound to RTKs, presumably due to the induced proximity. In the case of PLCγ1, tyrosine phosphorylation may be a requirement for activation of the enzyme (Nishibe et al., 1990). However, despite the high affinity interactions between PLCγ and pp140^{trk} (Ohmichi et al., 1991), and the NGF-dependent phosphorylation of PLCγ1 (Kim et al., 1991; Ohmichi et al., 1991), no significant increases in IP$_3$ or calcium mobilization are detected (Landreth et al., 1980). This may merely reflect the transient nature of this activation, but also raises the possibility that such RTK-SH2 protein interactions play separate regulatory roles, perhaps serving as an anchor for other proteins, such as the 38 kd associated phosphoprotein (Ohmichi et al., 1992b). Alternatively, some SH2 domains may modulate the RTK interactions of other phosphotyrosine binding proteins such as PTPases (Yi et al., 1992).

Specificity in Signal Initiation Is Produced by Combinatorial Diversity of Protein Interactions

The emerging large numbers of RTKs and SH2 proteins begs one central question: where is the specificity? Indeed, most SH2 domains can interact with phosphotyrosine alone, but at significantly higher concentrations than are required for RTK interaction. However, immunoprecipitation experiments have indicated that these molecules discriminate in their binding to receptors or substrates, perhaps providing the biochemical basis for initiation of the divergent cellular effects of growth factors. In the case of some receptors, typified by that for insulin and IGF-1, and intracellular kinases, such as the janus tyrosine kinase family that are activated by cytokine receptors, the tyrosine kinase can phosphorylate endogenous substrates that serve as surrogates for SH2 binding. The phosphorylation of IRS-1 by insulin or IL-4 induces stable complexes between IRS-1 and SH2 proteins, including the p85 regulatory subunit of PI-3 kinase (Becker et al., 1992), Grb2 (Myers et al., 1994), crk (Beitner-Johnson et al., 1995), SHPTP2 (Sugimoto et al., 1994), and others (Sun et al., 1993). These interactions can lead to the activation and/or localization of signaling cascades, including PI-3 kinase, mitogen-activated protein (MAP) kinase and others. These examples of different pathways of RTK-SH2 protein coupling may reflect a combinatorial diversity that allows for subtle differences in signal discrimination (Figure 2).

SERINE/THREONINE PHOSPHORYLATION CASCADES

Growth Factors Activate the MAP Kinase Pathway

In cells treated with growth factors or transfected with tyrosine kinase oncogenes, tyrosine phosphorylations are relatively scarce compared to those found on serine and threonine residues. These observations have led many investigators to suspect that serine kinases were activated directly or indirectly after tyrosine phosphorylation by growth factor receptors, thus dramatically amplifying the initial signal in the cell. Numerous growth factor-dependent protein kinases have been described. The best characterized of these is the family of enzymes known collectively as MAP kinases (Ferrell, 1986). These enzymes, which are stimulated by virtually every mitogen known, are themselves activated by phosphorylation on tyrosine and threonine residues (Miyasaka et al., 1990; Rossomando et al., 1991). Initially it was thought that this phosphorylation resulted from the direct actions of RTKs, although we now know that MAP kinases are activated by a complex series of protein:protein interactions and upstream protein kinases and phosphatases.

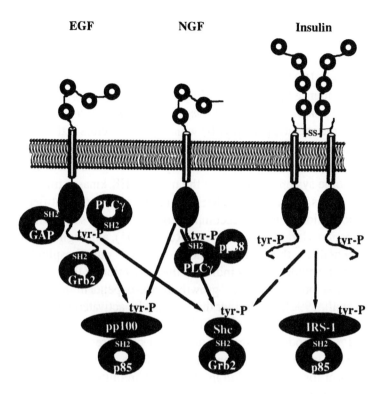

Figure 2. Combinatorial diversity in growth factor signaling. The specificity of signal initiation in growth factor action depends on the molecular interaction of receptors. Interactions of the EGF, NGF and insulin receptors with proteins containing *src* homology (SH) 2 domains are depicted. These three growth factors produce different effects on cells, but can activate some of the same pathways. The EGF receptor can form high affinity complexes with signaling proteins such as PLCγ1, *ras* GAP, and Grb2, through their respective SH2 domains. The pp140[*trk*] NGF receptor can interact with PLCγ1 and an associated 38 kd tyrosine phosphorylated protein, the function of which is unknown at present. In contrast, the insulin receptor does not form stable complexes with these SH2 domain proteins. All three of these growth factors stimulate the activity of PI-3 kinase, probably by the tyrosine phosphorylation of proteins that bind to the SH2 domain of the 85 kd regulatory subunit of the enzyme. In the case of the insulin receptor, this surrogate phosphoprotein, called IRS-1, has 10 known phosphorylation sites within the YMXM motif thought to be specifically recognized by p85. The mechanism of activation of PI kinase by EGF and NGF is less clear, but may involve phosphorylation of a 100 kd protein functionally analogous to IRS-1.

MAP kinase is directly activated by MAP kinase kinase or MEK (Crews et al., 1992; Shirakabe et al., 1992), a dual specificity kinase exhibiting sequence homology to the byrl and STE7 gene products of *Sacchromyces pombe* and *Sacchromyces cerevisiae* (Crews et al., 1992). Although the specific pathways by which growth factors activate this enzyme remain unknown, activation of the *raf* protooncogene (Rapp, 1991) is one of the predominant mechanisms by which MEK is activated. However, in some systems c-*raf* is not the major MEK kinase stimulated by growth factors, and additional MEK kinases have been identified.

Raf activation is usually accompanied by enhanced serine and threonine phosphorylation of the enzyme (Rapp, 1991), although it is still unclear whether this phenomenon reflects autophosphorylation. Considerable evidence now indicates that activation of the *ras* protooncogene is a prerequisite for growth factor-dependent *raf* activation (Wood et al., 1992), and further that *raf* may be activated via a direct interaction with *ras* (Zhang et al., 1993). Microinjection of neutralizing anti-p21[ras] antibodies (Hagag et al., 1986) or expression of dominant negative *ras* mutants (Szebereny et al., 1990) have revealed an absolute requirement for *ras* activation in the serine/threonine phosphorylations induced by many growth factors (Nakafuku et al., 1992).

ras Functions as a Molecular Switch in Regulating the MAP Kinase Pathway

Significant progress has been made in dissecting the molecular events involved in the regulation of p21[ras] by tyrosine kinase receptors. Activation appears to involve exchange of GDP by GTP on the *ras* protein, catalyzed by a nucleotide exchange factor. In most mammalian cells, the growth factor-sensitive exchange factor is presumed to be a homologue of the yeast protein *son of sevenless* (SOS) (Bowtell et al., 1992), although it is possible that other exchange factors exist. The increased binding of GTP to *ras* catalyzed by SOS is likely to be activated by the targeting of SOS to *ras* via the interaction with the adapter protein Grb2, mediated by the binding of one of the SH3 domains of Grb2 with a proline-rich region in the carboxy terminus of SOS. This Grb2-SOS interaction is itself thought to be stimulated by the binding of Grb2 to tyrosine phosphorylated proteins through its SH2 domain. The growth factor-induced phosphotyrosine-SH2 interaction may occur by direct binding to an autophosphorylated receptor, as is the case for the EGF receptor (Rozakis-Adcock et al., 1993), or may occur through a surrogate phosphorylated protein, such as shc or IRS-1 (Myers et al., 1994).

Recent evidence suggests that *ras* may require another signal for activation, possibly including the stimulation of protein dephosphorylation. Overexpression of a catalytically inactive SHPTP2 or Syp in NIH3T3 cells blocked the activation of the MAP kinase pathway by insulin (Milarski and Saltiel, 1994). As described above, the catalytic activity of this tyrosine phosphatase is increased upon the engagement of its SH2 domains with tyrosine phosphorylated proteins (Ploskey et al., 1995). Although the precise substrates of this enzyme have not been identified, Shc and IRS-1 do not undergo dephosphorylation upon Syp activation, suggesting that the impact of this phosphatase on *ras* depends upon a separate, but required, pathway (Noguchi et al., 1994).

The MAP Kinase Pathway Plays a Critical, but Limited, Role in Signal Transduction

Upon its activation, MAP kinase can translocate into the nucleus, where it catalyzes the phosphorylation of transcription factors such as $p62^{TCF}$, initiating a transcriptional program that leads the cell to commit to a proliferative or differentiative cycle (Figure 3). MAP kinase can also phosphorylate a number of other proteins involved in cellular signaling, including other kinases and phospholipases. The role of the MAP kinase pathway in cell growth and differentiation has been explored with pharmacological tools, as well as by the expression of dominant interfering mutant proteins in the pathway. Overexpression of mutant forms of MEK in which the two *raf* phosphorylation sites have been substituted with alanine can produce a dominant negative phenotype in some (Pajes et al., 1994), but not all cell types (Syu et al., 1996). The discovery of a specific inhibitor of MEK, PD98059, has added much to our understanding of this pathway (Dudley et al., 1995). This compound blocks growth factor-dependent activation of MAP kinase, inhibits DNA synthesis in a variety of cell types, and prevents neurotrophin-induced differentiation of PC-12 cells. PD98059 can revert *ras* transformed fibroblasts to a normal phenotype, suggesting that blockade of this pathway might be useful in preventing the progression of tumorigenesis (Dudley et al., 1995). In contrast to the profound effects of the MEK inhibitor on cellular growth, the compound was without effect on metabolism, particularly regarding the anabolic effects of insulin (Lazar et al., 1995). Thus, although numer-

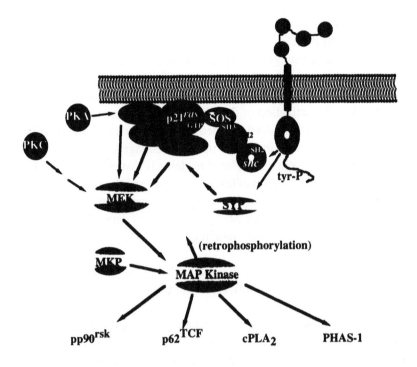

Figure 3. Regulation of the MAP kinase pathway. MAP kinases are activated by a variety of growth and differentiative signals. Tyrosine kinase receptors are thought to stimulate this pathway primarily through the activation of the *ras* protooncogene. Activation of growth factor receptors leads to the tyrosine phosphorylation of Shc, which then binds to Grb2 via an SH2 interaction. Shc can target the Grb2/SOS complex to the membrane, where SOS can stimulate the exchange of GDP for GTP on *ras*, thus switching this protein to the active state. Once activated, *ras* can recruit to the membrane protein kinases such as *raf* that can phosphorylate and activate the kinase MEK. Additional signals involving the tyrosine phosphatase Syp are also required in some cases for activation of *raf*. MEK is a dual specificity kinase that phosphorylates MAP kinase on threonine and tyrosine residues, leading to its activation. Upon activation, MAP kinase can phosphorylate a variety of substrates, including other kinases, such as p90[rsk], phospholipases, regulators of translation, such as PHAS1, and numerous transcription factors, including p62[TCF] and others.

ous hormonal agents can activate MAP kinase in different cell types, the pathway is not universally utilized in signal transduction, even for responses produced by some tyrosine kinase receptors.

In addition to its antiproliferative effects (Dudley et al., 1995), PD98059 also prevents the MAP kinase-mediated phosphorylation of other cytoplasmic substrates in some cells, including other protein kinases, phospholipase A2, and even the upstream signaling proteins known to be targets of desensitization. In this regard, PD98059 blocked the growth factor-induced phosphorylation of SOS, which causes this *ras* exchange factor to functionally uncouple from Grb2, attenuating the further activation of *ras* (Holt et al., 1996). Thus, in normal signaling paradigms, the activation of the MAP kinase pathway leads to a desensitizing retrophosphorylation that prevents the persistent activation of the *ras* protein, perhaps explaining in part why growth factor activation does not generally lead to oncogenesis.

In addition to the feedback regulation of signals leading to MAP kinase activation, the pathway may also be self limiting regarding the cellular consequences of activation. In the case of neuronal differentiation, the kinetics of MAP kinase activation are thought to play a major role in determining whether a cell commits to differentiate. In general, the prolonged activation and nuclear translocation of MAP kinase is associated with neurite outgrowth in PC-12 cells; other mitogenic stimuli that activate MAP kinase transiently do not produce neuronal differentiation (Decker, 1996). Expression of the trkA nerve growth factor (NGF) receptor in 3T3 cells leads to NGF-dependent growth arrest and differentiation, which is associated with induction of the cyclin-dependent kinase inhibitor p21$^{cip1/Waf1}$, and the subsequent down regulation of CDK4 activity (Pumiglia and Decker, 1996). These effects can be blocked by MEK inhibition, suggesting that the sustained activation of MAP kinase, produced by NGF, leads to the phosphorylation of proteins that are perhaps the products of early immediate genes, changing dramatically the phenotype of the cell (Pumiglia and Decker, 1996).

CONCLUSIONS

Dramatic advances have been made in our understanding of the molecular mechanisms involved in growth factor signaling. It appears that many of the important components in signaling pathways have been identified, and studies on their structure and function are likely to produce a fairly detailed picture of what role they play in governing cell growth, differentiation and survival. However, we still have much to learn regarding the molecular mechanisms that insure specificity in signal transduction, especially regarding the temporal and spatial relationships of different pathways to each other.

REFERENCES

Anderson, D., Koch, C.A., Grey, L., Ellis, C., Moran, M.F., & Pawson, T. (1990). Binding of SH2 domains of phospholipase Cγl GAP, and Src to activated growth factor receptors. Science 250, 979-982.

Becker, J.M., Myers, M.G., Shoelson, S.E., Chin, D.J., Sun, X.J., Miralpeix, M., Hu, P., Margolis, B., Skolnik, E.Y., Schlessinger, J., & White, M.F. (1992). Phosphatidylinositol-3 kinase is activated by association with IRS-1 during insulin stimulation. EMBO J. 11, 3469-3479.

Beitner-Johnson, D., & LeRoith, D. (1995). Insulin-like growth factor-1 stimulates tyrosine phosphorylation of endogenous c-crk. J. Biol. Chem. 270, 5187-5190.

Bowtell, D., Fu, P., Simon, M., Senior, P. (1992). Identification of murine homologues of the *Drosophila son of sevenless* gene: Potential activators of Ras. Proc. Natl. Acad. Sci. USA 89, 6511-6515.

Crews, C.M., Alessandrini, A., & Erikson, R. (1992). The primary structure of MEK, a protein kinase that phosphorylates the ERK gene product. Science 258, 478-480.

Decker, S.J. (1996). Nerve growth factor-induced growth arrest and induction of p21[Cip1/WAF1] in NIH-3T3 cells expressing TrkA. J. Biol. Chem. 270, 30841-30844.

Downing, J.R., Margolis, B., Zilberstein, A., Ashmum, R.A., Ullrich, A., Sherr, C.J., & Schlessinger, J. (1989). Phospholipase Cγ, a substrate for PDGF receptor kinase, is not phosphorylated on tyrosine during the mitogenic response to CSF-1. EMBO J. 8, 3345-3350.

Dudley, D.T., Pang, L., Decker, S.J., Bridges, A.J., & Saltiel, A.R. (1995). A synthetic inhibitor of the mitogen-activated protein kinase cascade. Proc. Natl. Acad. Sci. USA 92, 7686-7689.

Escobedo, J.A., Navankasattusas, S., Kavanaugh, W.M., Milfay, D., Fried, V.A., & Williams, L.T. (1991). cDNA cloning of a novel 85 Kda protein that has SH2 domains and regulates binding of PI3-kinase to the PDGF beta-receptor. Cell 65, 75-82.

Ferrell, J.E. (1986). MAP kinase in mitogenesis and development. Curr. Topics Dev. Biol. 33, 1-60.

Hagag, N., Halegoua, S., & Viola, M. (1986). Inhibition of growth factor-induced differentiation of PC-12 cells by microinjection of antibody to Ras p21. Nature 319, 680-682.

Holt, K.H., Waters, S.B., Okada, S., Yamauchi, K., Decker, S.J., Saltiel, A.R., Motto, D.G., Koretsky, G.A., & Pessin, J.E. (1996). Epidermal growth factor receptor targeting prevents uncoupling of the Grb2-SOS complex. J. Biol. Chem. 271, 8300-8306.

Hu, P., Margolis, B., Skolnik, E.Y., Lammers, R., Ullrich, A., & Schlessinger, J. (1992). Interaction of phosphatidylinositol-3-kinase-associated p85 with EGF and PDGF receptor. Mol. Cell. Biol. 12, 981-990.

Kavanaugh, W.M., Turck, C.W., & Williams, L.T. (1995). PTB domain binding to signaling proteins through a sequence motif containing phosphotyrosine. Science 268, 1177-1179.

Kim, U.H., Fink, J.D., Kim, H.S., Park, D.J., Contreras, M.L., Guroff, G., & Rhee, S.G. (1991) Nerve growth factor stimulates phosphorylation of phospholipase Cγ in PC-12 cells. J. Biol. Chem. 266, 1359-1362.

Koch, C.A., Anderson, D., Moran, M.F., Ellis, C., & Pawson, T. (1991). SH2 and SH3 domains: Elements that control interactions of cytoplasmic signaling proteins. Science 252, 668-674.

Kosako, H., Gotoh, Y., Matsuda, S., Ishikawa, M., Nishida, E. (1992). Xenopus MAP kinase activator is a serine/threonine/tyrosine kinase activated by threonine phosphorylation. EMBO J. 11, 2903-2908.

Kuriyan, J., & Cowburn, D. (1993). Structures of SH2 and SH3 domains. Current Opin. Struct. Biol. 3, 828-837.

Landreth, G.E., Cohen, P., & Shooter, E.M. (1980). Ca^{2+} transmembrane fluxes and nerve growth factor action on a clonal cell line of rat pheochromocytoma. Nature 283, 202-204.

Lazar, D.F., Brady, M.J., Wiese, R.J., Mastick, C.C., Waters, S.B., Yamauchi, K., Pessin, J.E., Cuatrecasas, P., & Saltiel, A.R. (1995). MEK inhibition does not block the stimulation of glucose utilization by insulin. J. Biol. Chem. 270, 20801-20807.

Mayer, B.J., Jackson, P.K., Etten, R.A.V., & Baltimore, D. (1992). Point mutations in the abl SH2 domain coordinately impair phosphotyrosine binding *in vitro* and transforming activity *in vivo*. Mol. Cell. Biol. 12, 609-618.

McNamara, D., Dobrusin, E., Zhu, G., Decker, S.J., & Saltiel, A.R. (1993). Inhibition of binding of phospholipase Cγ1 SH2 domains to phosphorylated epidermal growth factor receptor by phosphorylated peptides. Intl. J. Peptide and Protein Res. 42, 240-248.

Milarski, K.L., & Saltiel, A.R. (1994). Expression of catalytically inactive SYP phosphatase in 3T3 cells blocks the stimulation of MAP kinase by insulin. J. Biol. Chem. 269, 21239-21243.

Miyasaka, T., Chao, M.V., Sherline, P., & Saltiel, A.R. (1990). Nerve growth factor stimulates a protein kinase in PC-12 cells that phosphorylates microtubule-associated protein-2. J. Biol. Chem. 265, 4730-4735.

Myers, M.G., Wang, L-M., Sun, X.J., Zhang, Y., Yenush, L., Schlessinger, J., Pierce, J.H., & White, M.F. (1994). Role of IRS-1-GRB-2 complexes in insulin signaling. Mol. Cell. Biol. 14, 3577-3587.

Nakafuku, M., Satoh, T., & Kaziro, Y. (1992). Differentiation factors, including nerve growth factor, fibroblast growth factor, and interleukin-1 induce an accumulation of an active Ras GRP complex in Rat pheochromocytoma PC-12 cells. J. Biol. Chem. 267, 19448-19454.

Nishibe, S., Wahl, M.I., Hernandex-Solomayor, S.M.T., Tonks, N.K., Rhee, S.G., & Carpenter, G. (1990). Increase of the catalytic activity of phospholipase Cγ1 by tyrosine phosphorylation. Science 250, 1253-1256.

Noguchi, T., Matozoki, T., Horita, K., Fujioka, T., & Kajuga, M. (1994). Role of SHPTP2, a protein tyrosine phosphatase with SH2 domains in insulin-stimulated ras activation. Mol. Cell. Biol. 14, 6674-6682.

Ohmichi, M., Decker, S.J., Pang, L., & Saltiel, A.R. (1991). Nerve growth factor binds to the 140 Kda Trk protooncogene product and stimulates its association with Src homology domain of phospholipase C-γ1. Biochem. Biophys. Res. Commun. 179, 217-223.

Ohmichi, M., Decker, S.J., & Saltiel, A.R. (1992a). Nerve growth factor stimulates the tyrosine phosphorylation of a 38 kDa protein that specifically associates with the Src homology domain of phospholipase Cγ1. J. Biol. Chem. 267, 21601-21606.

Ohmichi, M., Decker, S.J., & Saltiel, A.R. (1992b). Activation of phosphatidylinositol-3-kinase by nerve growth factor involves indirect coupling of the Trk protooncogene with Src homology 2 domains. Neuron 9, 767-777.

Overduin, M., Rios, C.B., Mayer, B.J., Baltimore, D., & Cowburn, D. (1992). Three dimensional solution structure of the Src homology 2 domain of c-abl. Cell 70, 697-704.

Pajes, G., Brurnet, A., L'Allemain, G., & Pouyssegur, J. (1994). Constructive mutant and putative regulatory serine phosphorylation site of mammalian MAP kinase kinase (MEK 1). EMBO J. 13, 3003-3010.

Ploskey, S., Wandless, T.J., Walsh, C.T., & Shoelson, S.E. (1995). Potent stimulation of SHPTP2 phosphatase activity by simultaneous occupancy of both SH2 domains. J. Biol. Chem. 270, 2897-2900.

Pumiglia, K., & Decker, S.J. (1996). Cell cycle arrest mediated by the MEK/MAP kinase pathway. Submitted.

Rapp, U.R. (1991). Role of Raf-1 serine/threonine kinase in growth factor signal transduction. Oncogene 6, 495-500.

Reedijk, M., Liu, X., & Pawson, T. (1990). Interactions of phosphatidylinositol kinase, GTPase-activating protein (GAP), and GAP-associated proteins with the colony-stimulating factor 1 receptor. Mol. Cell. Biol. 10, 5601-5608.

Rossomando, A.J., Rayne, D.M., Weber, M.J., Sturgill, T.W. (1991). Evidence that pp42, a major tyrosine target protein is a mitogen-activated serine/threonine protein kinase. Proc. Natl. Acad. Sci. USA 86. 6940-6943.

Rozakis-Adcock, M., Fernley, X., Wade, J., Pawson, T., & Bowtell, D. (1993). The SH2 and SH3 domains of mammalian Grb2 couple the ECR receptor to the ras activator mSOS1. Nature 363, 83-85.

Schlessinger, J., & Ullrich, A. (1992). Growth factor signalling by receptor tyrosine kinases. Neuron 9, 383-391.

Shirakabe, K., Gotoh, Y., & Nishida, E. (1992). A mitogen-activated protein (MAP) kinase activating factor in mammalian mitogen-stimulated cells in homoogous to Xenopus M phase MAP kinase. J. Biol. Chem. 267, 16685-16690.

Songyang, Z., Shoelson, S.E., Chaudhuri, M., Gish, G, Pawson, T., Haser, W.G., King, F, Roberts, T., Retnofsky, S., Lechleider, R.J., Neel, B.G., Birge, R.B., Fajardo, J.E., Cou, M.M., Hanfusa, H., Schaffhausen, B, & Cantley, L.C. (1993). SH2 domains recognize specific phosphopeptide sequences. Cell 72, 767-778.

Sugimoto, S. Wandless, T.J., Shoelson, S.E., Neel, B.G., & Walsh, C.T. (1994). Activation of the SH2-containing peptides derived from insulin receptor substrate-1. J. Biol. Chem. 269, 13614-13622.

Sun, X.J., Crimmins, D.L., Myers, Jr., M.G., Miralpeix, M., & White, M.F. (1993). Pleiotropic insulin signals are engaged by multisite phosphorylation of IRS-1. Mol. Cell. Biol. 13, 7418-7426.

Syu, L-J., Guan, K., & Saltiel, A.R. (1996). Submitted.

Szebereny, J., Cai, H., & Cooper, G.M. (1990). Effect of a dominant inhibitory Ha-ras mutation on neuronal differentiation of PC-12 cells. Mol. Cell. Biol. 10, 5324-5332.

van der Geer, P., & Pawson, T. (1995). The PTB domain: a new protein module implicated in signal transduction. TIBS 20, 277-280.

Wood, K.M., Sarnecki, C., Roberts, T.M., Blenis, J. (1992). Ras mediates nerve growth factor receptor modulation of three signal transducing protein kinases: MAP kinase, Raf-1, and RSK. Cell 68, 1041-1050.

Yi, T., Cleveland, J.L., & Ihle, J.N. (1992). Protein tyrosine phosphatase containing SH2 domains: Characterization, preferential expression in hematopoietic cells, and localization to human chromosome 12p12-p13. Mol. Cell. Biol. 12, 836-846.

Zhang, X-F, Settleman, J., Kyriakis, J.M., Takeuchi-Suzuki, E., Elledge, S.J., Marshall, M.S., Bruder, J.T., Rapp, U.R., & Avruch, J. (1993). Normal and oncogenic p21^{ras} proteins bind to the amino terminal regulatory domain of c-raf-1. Nature 364, 308-313.

Zhou, M-M., Ravichandran, K.S., Olejniczak, E.T., Petros, A.M., Meadows, R.P., Settler, M., Harlan, J.E., Wade, W.S., Burakoff, S.J., & Fesik, S.W. (1995). Structure and ligand recognition of the phosphotyrosine binding domain of Shc. Nature 378, 584-592.

Zhu, G., Decker, S.J., & Saltiel, A.R. (1992). Direct analysis of the binding of Src-homology 2 domains of phospholipase C to the activated epidermal growth factor receptor. Proc. Natl. Acad. Sci. USA 89, 9559-9563.

Zhu, G., Decker, S.J., Mayer, B.J., & Saltiel, A.R. (1993). Direct analysis of the binding of the abl Src homology 2 domain to the activated epidermal growth factor receptor. J. Biol. Chem. 266, 12964-12970.

Chapter 5

Tissue-Specific Expression of the *CYP19* (Aromatase) Gene

EVAN R. SIMPSON, M. DODSON MICHAEL,

VEENA R. AGARWAL, MARGARET M.

HINSHELWOOD, SERDAR E. BULUN,

and YING ZHAO

Advances in Molecular and Cellular Endocrinology
Volume 2, pages 99-120.
Copyright © 1998 by JAI Press Inc.
All rights of reproduction in any form reserved.
ISBN: 0-7623-0292-5

INTRODUCTION

Aromatase (Thompson, et al., 1974; Mendelson et al., 1985; Nakajin et al., 1986; Kellis and Vickery, 1987) (P450arom, the product of the *CYP19* gene (Nelson et al., 1993)) is the enzyme which catalyses the biosynthesis of estrogens. *CYP19* is a member of the P450 superfamily of genes which currently contains over 300 members in some 36 gene families (Nelson et al., 1993). Estrogen biosynthesis appears to occur throughout the entire vertebrate phylum including mammals, birds, reptiles, amphibians, teleost and elasmobranch fish, and *Agnatha* (hagfish and lampreys) (Callard et al., 1900, 1980; Callard, 1981). It has also been described in the protochordate *Amphioxus* (Callard et al., 1984). To our knowledge, estrogen biosynthesis has not been reported in non-chordate animal phyla. This is in spite of the fact that the CYP19 family appears to be an ancient lineage of P450 gene products, diverging as much as 10^9 years ago (Nelson et al., 1993). In most vertebrate species that have been examined, aromatase expression occurs in the gonads and in the brain. This is true of the fish and avian species that have been examined as well as most mammals (such as rodents). In many species, estrogen biosynthesis in the brain has been implicated in sex-related behavior such as mating responses, and frequently a marked sexually dimorphic difference has been demonstrated. This is true, for example, in avian species in which the song of the male is important in courtship behavior (Hutchinson, 1991). In the case of humans and a number of higher primates, there is a more extensive tissue distribution of estrogen biosynthesis, since this also occurs in the placenta of the developing fetus as well as in the adipose tissue of the adult. The ability of the placenta to synthesize estrogen is also the property of a number of ungulate species such as cows, pigs, and horses. However, at least in cattle, there is no evidence of estrogen biosynthetic capacity in adipose, whereas in rodent species such as rat and mice, as well as in rabbits, neither adipose nor placenta has any ability to synthesize estrogens.

The physiological significance of estrogen biosynthesis in the placenta and adipose of humans is unclear at this time. The C_{18} steroid produced in each tissue site of biosynthesis is quite tissue-specific. For example, the human ovary synthesizes primarily estradiol, whereas the placenta synthesizes estriol and adipose synthesizes estrone. This appears to reflect primarily the nature of the C_{19} steroid presented to the estrogen-synthesizing enzyme in each tissue site. Thus, in the case of adipose tissue, the principal source of substrate is circulating androstenedione produced by the adrenal cortex. In the case of the placenta, the major precursor is 16α-hydroxydehydroisoandrosterone sulfate derived as a consequence of the combined activities of the fetal adrenal and liver. Although the human placenta produces very large quantities of C_{18} steroids, particularly estriol, its physiological importance is unclear. This pertains because in pregnancies characterized by placental sulfatase deficiency, the placenta is essentially deprived of C_{19} substrate and hence synthesizes, relatively speaking, minute quantities of estrogen, yet such pregnancies are quite uncomplicated (France and Liggins, 1969). At most, parturition is delayed by several days. Similarly, at this time no physiological significance has been attributed to estrogen biosynthesis by human adipose; however, the latter has been implicated in a number of pathophysiological conditions. Estrogen biosynthesis by adipose tissue not only increases as a function of body weight but as a function of age (Hemsell et al., 1974; Edman and MacDonald, 1976) and has been correlated directly with the incidence of endometrial cancer as well as with postmenopausal breast cancer. Furthermore, evidence is accumulating to suggest that the estrogen implicated in the development of breast cancer is that which is produced locally within the adipose tissue of the breast itself (Miller and O'Neill, 1987). On the other hand, estrogen biosynthesis in adipose tissue may have beneficial consequences since osteoporosis is more common in small, thin women than in large, obese women. While this may be, in part, the consequence of the bones of the latter being subject to load-bearing exercise, nonetheless it seems likely that the increased production of estrogens by the adipose of obese women is a significant factor.

THE *CYP19* GENE

Some years ago we and others cloned and characterized the *CYP19* gene which encodes human P450arom (Means et al., 1989; Harada et al., 1990; Toda et al., 1990) (Figure 1). The coding region spans nine exons beginning with exon II (Figure1). Sequencing of rapid amplification of cDNA ends (RACE)-generated cDNA clones derived from P450arom transcripts present in the various tissue sites of expression revealed that the 5'-termini of these transcripts differ from one another in a tissue-specific fashion upstream of a common site in the 5'-untranslated region

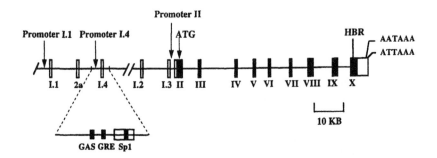

Figure 1. Schematic representation of the human *CYP19* gene. The closed bars represent translated sequences. The septum in the open bar in exon II represents the 3'-acceptor splice junction for the untranslated exons. The sequence immediately to the left of the septum is that present in mature transcripts whose expression is driven by promoter II. The five untranslated exons, I.1, 2a, I.2, I.3, and I.4, are indicated in their approximate locations. Also shown are promoters I.1 and II and putative promoters I.2, I.3, and I.4. The heme-binding region (HBR) is indicated in exon X, as are two alternative polyadenylation signals that give rise to the two species of P450arom transcript of 3.4 and 2.9 kb. The genomic region shown spans a distance of at least 75 kb; however, as the gap between exons I.4 and I.2 has never been bridged, the true size is unknown.

(Means et al., 1991; Kilgore et al., 1992; Jenkins et al., 1993; Toda and Shizuta, 1993). Using these sequences as probes to screen genomic libraries, it was found that these 5'-termini correspond to untranslated exons which are spliced into the P450arom transcripts in a tissue-specific fashion, due to the use of tissue-specific promoters. Placental transcripts contain at their 5'-ends untranslated exon I.1 which is located at least 40 kb upstream from the start of translation in exon II (Mahendroo et al., 1991; Means et al., 1991). This is because placental expression is driven from a powerful distal placental promoter, I.1, upstream of untranslated exon I.1. On the other hand, transcripts in the ovary contain sequence at their 5'-ends which is immediately upstream of the start of translation. This is because expression of the gene in the ovary utilizes a proximal promoter, promoter II (Jenkins et al., 1993). By contrast, transcripts in adipose tissue contain yet another distal untranslated exon, I.4, which is located in the gene 20 kb downstream from exon I.1 (Mahendroo et al., 1993). A number of other untranslated exons have been characterized by ourselves and others (Harada et al., 1993; Toda et al., 1994), including one specific for brain (Honda et al., 1994). Splicing of these untranslated exons to form the mature transcripts occurs at a common 3'-splice junction which is upstream of the start of translation. This means that although transcripts in different tissues have different 5'-termini, the protein encoded by these transcripts is always the same, regardless of the tissue-site of expression, thus there is only one human P450arom enzyme encoded by a single-copy gene.

EXPRESSION OF AROMATASE IN HUMAN OVARY

As indicated previously, aromatase P450 is expressed in the pre-ovulatory follicles and corpora lutea of ovulatory women by means of a promoter proximal to the start of translation (PII) (Means et al., 1991; Jenkins et al., 1993). Aromatase expression in the granulosa cells of the ovary is primarily under the control of the gonadotropin follicle-stimulating hormone (FSH), whose action is mediated by cAMP. To understand how this transcription is controlled by cAMP, we constructed chimeric constructs containing deletion mutations of the proximal promoter 5'-flanking DNA fused to the rabbit β-globin reporter gene (Michael et al., 1994). Assay of reporter gene transcription in transfected bovine granulosa and luteal cells revealed that basal and cAMP-stimulated transcription was lost upon deletion from -278 to -100 bp, indicating the presence of a functional response element in this region; however, no classical cAMP-responsive element was found. Mutation of a CAAGGTCA motif located at -130 bp revealed that this element is crucial for basal and cAMP-stimulated reporter gene transcription. When a single copy of this element was placed upstream of a heterologous promoter, it could act as a weak cAMP-response element (CRE). Electrophoretic mobility shift assay in the presence of specific antibodies and u/v-cross-linking established that Ad4BP/SF-1 binds to this hexameric element. Steroidogenic factor-1 (SF-1) is an orphan member of the steroid hormone receptor gene superfamily which has been shown to be a critical developmental factor for the gonads as well as the adrenals (Lala et al., 1992; Luo et al., 1994). However deletion mutation analysis revealed the presence of another sequence upstream of the SF-1 site which was also critical for cAMP-responsiveness. This sequence, at -211/-202 bp, is TGCACGTCA, identical to a canonical CRE save for the extra C. A similar element is present in the rat gene (Fitzpatrick and Richards, 1994). In the case of the human, electrophoretic gel mobility shift and antibody super-shift analysis revealed that a number of factors bind to this site (Michael and Simpson, 1996). Moreover, u/v-cross-linking analysis revealed that the major protein binding to this site appeared to be about 24 kDa in size, much smaller than CRE binding protein (CREB) or, indeed, most other transcription factors. Thus, it is apparent that much remains to be elucidated regarding the mechanism whereby FSH and cAMP regulate aromatase expression in the ovary.

EXPRESSION OF AROMATASE IN HUMAN PLACENTA

As indicated previously, placental expression of aromatase in the human is driven from a powerful distal placental promoter I.1 upstream of untranslated exon I.1 which is located at least 40 kb upstream from the start of translation in exon II (Means et al., 1991). At this time, the true distance is unknown since the genomic clones containing exon I.1 on the one hand and exon II on the other have never been overlapped. Employing various deletion mutations of the upstream flanking region of exon I.1, we and others have examined several putative regulatory sequences within this region and the proteins which interact with these sequences to regulate expression of aromatase in choriocarcinoma cells. Thus Toda and colleagues (1992) have identified a binding site for C/EBP-β which is located between -2141 and -2115 bp relative to the start of transcription in exon I.1. They further identified an element located between -238 and -200 bp which appears to synergize with the C/EBP-β element upstream. Yamada *et al*, (1995) identified two elements within -300 bp upstream of exon I.1 which recognize the same trans-acting factor that binds to the trophoblast-specific element previously located in the enhancer region of the human glycoprotein hormone α-subunit gene.

We have identified an imperfect palindromic sequence 5'-AGGTCATGCCCC-3' located at -183 to -172 bp which is responsible for stimulation of aromatase expression by retinoic acids (Sun et al., 1996). This does not function as a binding site for SF-1 since SF-1 is not expressed in placenta or in the cells, however, it does appear to bind a heterodimer comprising retinoid X receptor α (RXRα) and vitamin D receptor (VDR). It was reported recently that levels of RXR and retinoic acid receptor (RAR receptor expression increased during the process of cytotrophoblast differentiation into syncytiotrophoblasts in the placenta (Stephanon et al., 1994). This is coincident with the increase in aromatase expression. These results suggests that retinoids may indeed play an important role in developmental regulation of aromatase gene expression in the placenta.

EXPRESSION OF AROMATASE IN ADIPOSE TISSUE

We have found that aromatase expression does not occur in adipocytes but rather in the stromal cells which surround the adipocytes, and which may themselves be preadipocytes (Price et al., 1992a). These stromal cells grow in culture as fibroblasts. Consequently we have employed these cells in primary culture as a model system to study the regulation of estrogen biosynthesis in adipose tissue (Ackerman et al., 1981). When serum is present in the culture medium, expression is stimulated by glucocorticoids including dexamethasone (Simpson et al., 1981). Under these conditions P450arom transcripts contain primarily untranslated exon I.4 at their 5'-end (Mahendroo et al., 1993; Zhao et al., 1995a). We subsequently have characterized the region of the *CYP19* gene upstream of exon I.4 (Figure 2) and have found it to contain a TATA-less promoter, as well as an upstream glucocorticoid response element (GRE) and an Sp1 sequence within the untranslated exon, both of which are required for expression of reporter gene constructs in the presence of serum and glucocorticoids (Zhao et al., 1995a). Additionally, we found this region to contain an interferon-γ activating sequence (GAS) element. Such sequences are known to bind transcription factors of the signal transducer and activator of transcription (STAT) (Schindler et al., 1992; Darnell, et al., 1994; Zhong et al., 1994).

We have studied aromatase expression in samples of adipose tissue obtained from women of various ages, using competitive reverse transcription-polymerase chain reaction (RT-PCR) with an internal standard, and have found a marked increase in the specific content of P450arom transcripts in adipose tissue with increasing age (Bulun and Simpson, 1994), thus providing a molecular basis for the previous observation that the fractional conversion of circulating androstenedione to estrone increases with age (Hemsell et al., 1974; Edman and MacDonald, 1976). Furthermore, there are marked regional variations in aromatase expression, with highest values being found in adipose from buttocks and thighs as compared with abdomen and breast (Killinger et al., 1987; Agarwal et al., 1997).

We also used this RT-PCR technique to examine regional variations in aromatase expression in breast adipose tissue and have found that the highest expression occurs in adipose tissue proximal to a tumor, as compared to that distal to a tumor (Bulun et al., 1993a; Agarwal et al., 1996). This is in agreement with previous observations regarding the regional distribution of aromatase activity within breast adipose (O'Neill et al., 1988; Reed et al., 1993), as well as an immunocytochemical study (Sasano et al., 1994). These results

```
ctctggtcag atattttgat catgctacag tgcatgaaat tgttcataag   -754

aattgtatgt gcctctgtat ctaacaggat ctgcttatat cttcagaaaa   -704

ctttgtcata aatttaaatt acttaaagtg tctgatcttc agatacttta   -654

aagtagtgca tttgagaatg ggaatgttga ttacagtgcg tatagggaaa   -604

tagatgaata ttccattaat aactattaaa atctgctaaa gcttaggcta   -554

agctgatata tttagttgta ataaaattgg gtgaacacat tccaacttca   -504

gcctgattaa gggaaagggt gtaggggtga gacacttagg cggagcttga   -454

aaaggaatgg tgagagtttg gccaatggaa ggaaggctgt gccagacagg   -404

aatagtgtgg gctgacgaca actgagggca aagtgcttgt cccctcatag   -354

ttgcgcaatg aatgcagagg ggctgaggtt catctgtcgt cttcagctct   -304
                              GAS
gcaggctaca tctcagggtg ┌ttcctgtga a┐gttccaga agaaagctgt    -254

atggtcagct tggggaaata tgtggttcat gctggaatgc tggacatacc   -204

acattattgg aaagatgcac attgaatgac cgacaaaatg aaactcaact   -154
                              GRE
ttccaaatgc tggtaatgag ┌agaagattct gttct┐aatga ccagttgttt   -104

cctgaaagaa tgtcagctcg attcataatg aatgcattct aaccatgaca    -54

gccacagtca ggacacaaaa aacaaagtgt ccttgatccc aggaaacagc     -4
        +1
cctCTGGAAT CTGTGTAAAT CTAGAAACAT AGTTGGGAAA ACTCTGACAC    +47

CCCTGCCCCA TGACCAACCA AGACTAAGAG TCCCAGAAGA TGGAGGTCAC    +97

AGAAGGCAGA GGCCTGCCCC CTCTCCAGGA GATCCCTGAC CCATGTGGGG   +147
              Sp1
TCAT┌GGGCGG GG┐CATGAGTG ATGTGATGGG AAACTGGCTC CTGGCTCCAA   +197

GTAGAACGTG ACCAACTGGA GCCTGACAGG AGAGTCCCTG GCACTGGTCA   +247

GCCCATCAAA CCAAG                                         +262
```

Figure 2. Sequence of the *CYP19* gene upstream of and including exon I.4. The start of transcription, determined by primer extension and S1 nuclease analysis, is indicated as +1. Consensus binding sites, namely GRE, GAS and Sp1, are shown in boxes. The nucleotides comprising exon I.4 are shown in capital letters.

suggest there is crosstalk between a breast tumor and the surrounding adipose cells in terms of the ability of the latter to synthesize estrogens, and that factors produced by developing breast tumors may set up local gradients of estrogen biosynthesis in the surrounding fat via paracrine and autocrine mechanisms(Simpson et al., 1994; Agarwal et al., 1996).

CYTOKINES WHICH STIMULATE AROMATASE EXPRESSION IN ADIPOSE TISSUE

Recently we observed for the first time that the effect of serum to stimulate aromatase expression in human adipose stromal cells (in the presence of glucocorticoids) can be mimicked by specific factors, namely members of the class I cytokine family, which includes interleukin-11 (IL-11), IL-6, oncostatin-M (OSM), and leukemia inhibitory factor (LIF) (Narazaki et al., 1994; Stahl et al., 1994). Members of this cytokine family employ a receptor system involving two different janus tyrosine kinase (JAK)-associated components, gp 130 and LIFRβ or a related β-component (Stahl and Yancopoulos, 1993). However, the IL-6 receptor complex includes a component whose cytoplasmic domain is apparently not involved in signaling (Stahl and Yancopoulos, 1993), and which can exist in a soluble form (Kishimoto et al., 1992). Recently an α-subunit of the IL-11 receptor complex has been cloned (Hilton et al., 1994), although this does not apparently exist in a soluble form. The concentration-dependence of the stimulation of aromatase by IL-6, IL-11, LIF, and OSM is indicative of high affinity receptor binding.

Addition of class I cytokines to adipose stromal cells resulted in a rapid phosphorylation of JAK1 (Zhao et al., 1995b). By contrast, JAK3 was not phosphorylated under these conditions to any significant extent, whereas JAK2 was phosphorylated to an equal extent both in the presence or absence of IL-11. As indicated by blotting with an anti-phosphotyrosine antibody and by inhibition in the presence of herbimycin A, this phosphorylation occurred on tyrosine residues present in the JAK1. Both gp130 and LIFRβ can associate with and activate at least three members of the JAK family, JAK1, JAK2, and TYK2, but utilize different combinations of these in different cells (Stahl et al., 1994), however, it is apparent that JAK1 is the kinase of choice in human adipose stromal cells.

This activation of JAK1 results in the rapid phosphorylation of STAT3 on tyrosine residues, but this was not the case for STAT1. Recently it has been shown that STAT3 is the substrate of choice for the IL-6/LIF/OSM cytokine receptor family, and that the specificity of STAT phosphorylation is based not upon which JAK is activated (Boulton et al., 1994; Stahl et al., 1994; Zhong et al., 1994), but rather is determined by specific tyrosine-based motifs in the receptor components, namely gp130 and LIFRβ, shared by these cytokines (Stahl et al., 1995). Finally, gel shift analysis indicated that STAT3 can interact with the GAS element present in the promoter I.4 region of the P450arom gene upon addition of IL-11 to these cells. This interaction in turn results in activation of expression, as indicated by transfection experiments employing chimeric constructs in which the region -330/+170 bp of the I.4 promoter region was fused upstream of the CAT reporter gene. The results indicate that both deletion of the GAS sequence, as well as mutagenesis of this sequence, resulted in complete loss of IL-11- and serum-stimulated expression in the presence of glucocorticoids.

Activation of this pathway of expression by these cytokines is absolutely dependent on the presence of glucocorticoids. This action of glucocorticoids is mediated by the GRE element downstream of the GAS element (Zhao et al., 1995a). Additionally, the Sp1-like element present within untranslated exon I.4 also is required, at least for expression of the -330/+170bp construct (Zhao et al., 1995a). These sequences, while present within a 400 bp region of the gene, are not contiguous and the nature of the interaction between STAT3, the glucocorticoid receptor and Sp1 to regulate expression of the P450arom gene via the distal promoter I.4 remains to be determined. Our present understanding of the regulation of expression of aromatase in adipose tissue is summarized in Figure 3.

Recently, we have found that tumor necrosis factor α (TNFα) also stimulates aromatase expression in adipose stromal cells in the presence of dexamethasone. This action of TNFα is mimicked by ceramide, indicative that sphingomyelinase activity is involved in the TNFα response. This action of TNFα appears to involve promoter I.4, specifically an AP1 site upstream of the GAS element which binds a c-jun/fos heterodimer upon activation by TNFα (Zhao et al., 1996b).

MESENCHYMAL - EPITHELIAL INTERACTIONS IN REGULATION OF AROMATASE EXPRESSION IN ADIPOSE TISSUE

As indicated previously, adipose tissue is the major site of estrogen biosynthesis in elderly women and men. The fact that this expression is confined to the stromal cells rather than the adipocytes themselves is consistent with the known actions of IL-6, IL-11, and TNFα to inhibit the differentiation of 3T3 L1 fibroblasts into adipocytes (Keller et al., 1993), thus aromatase is a marker for the undifferentiated fibroblast state. As indicated previously, aromatase expression in adipose increases dramatically with age (Keller et al., 1993; Bulun and Simpson, 1994). There is also a marked regional distribution with expression being greatest in buttock and thigh regions as compared to abdomen and breast (Price et al., 1992b; Bulun and Simpson, 1994). However, within the breast there is also a marked regional variation with expression being highest at sites

AROMATASE EXPRESSION IN ADIPOSE STROMAL CELLS

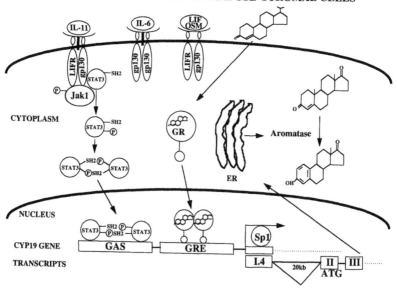

Figure 3. Schematic of second-messenger signaling pathways whereby Group I cytokines stimulate aromatase gene expression in human adipose stromal cells. JAK1 may be bound to the common receptor subunit gp 130 and activated following ligand binding and receptor dimerization, as a consequence of phosphorylation on tyrosine residues. STAT3 is recruited to binding sites on gp130 and is phosphorylated on tyrosine residues by JAK1. These phosphotyrosine residues are recognized by SH2-homology domains on STAT3, resulting in dimerization followed by translocation to the nucleus and binding to the GAS element of promoter I.4 of the aromatase gene. Following binding of glucocorticoid receptors to the GRE and Sp1 to its site on untranslated exon I.4, activation of transcription of the aromatase gene from promoter I.4 is initiated. Splicing of the initial transcript results in formation of mature mRNA which translocates to the ribosomes and is translated to give rise to aromatase protein.

proximal to a tumor as compared to those distal to a tumor (Bulun et al., 1993a; Agarwal et al., 1996).

Recently we developed a competitive RT-PCR technique to measure the levels of the various P450arom transcripts in adipose tissue. We found that in breast, abdomen, buttocks, and thighs of healthy subjects, I.4-containing transcripts predominated, with I.3- and II-specific transcripts in much lower abundance. Based on these findings we suggest that aromatase expression in adipose tissue may be under tonic control by circulating glucocorticoids and that regional and age-dependent variations may be the consequence of paracrine and autocrine secretion of stimulatory cytokines such as IL-6 and TNFα, the levels of which have been shown to increase with age (Wei et al., 1992; Daynes et al., 1993).

To our surprise, however, we found that the increase in aromatase expression in a tumor-containing breast was not due to an increase in I.4-specific transcripts, but rather of transcripts specific for promoters II and I.3 (Agarwal et al., 1996). Similar results have been obtained by Harada and colleagues (Harada et al., 1993). Since expression from these promoters is regulated by cAMP, these results strongly suggest that breast tumors secrete a factor(s) which stimulates aromatase expression in the surrounding stroma by increasing adenylate cyclase. Our recent evidence suggests that this factor is prostaglandin E_2 (PGE$_2$) (Zhao et al., 1996a). PGE$_2$ is a powerful stimulator of aromatase expression via both the protein kinase A (PKA) and protein kinase C (PKC) pathways. Moreover PGE$_2$ is known to be produced by breast tumor fibroblasts and epithelium, as well as by macrophages recruited to the tumor site (Schrey and Patel, 1995).

Such local paracrine mechanisms could be important in the stimulation of breast cancer growth by estrogens. Commonly, breast tumors produce a desmoplastic reaction whereby there is local proliferation of stromal cells surrounding the tumor, strongly indicative of the production of growth factors by the tumor. These proliferating stromal cells express aromatase, as indicated by immunocytochemistry (Sasano et al., 1994). It is possible then to propose a positive feedback loop whereby adipose stromal cells surrounding a developing tumor produce estrogens which stimulate the tumor to produce prostanoids, growth factors and cytokines (Dickson and Lippman, 1987). Some of these act to stimulate the further growth and development of the tumor in a paracrine and autocrine fashion. Additionally, these or other factors act to stimulate proliferation of the surrounding stromal cells and expression of aromatase within these cells. Thus a positive feedback loop is established by paracrine and autocrine mechanisms which lead to the continuing growth and development of the tumor.

AROMATASE EXPRESSION IN OTHER TISSUES

Stimulation of aromatase expression by serum in the presence of glucocorticoids is not confined to cells present in adipose tissue, but also has been reported in skin fibroblasts (Berkovitz et al., 1989) and in hepatocytes derived from fetal liver (Lanoux et al., 1985). In each of these cell types the P450arom transcripts contain exon I.4 as their 5'-terminus (Harada, 1992; Toda et al., 1994); however, the factors which mimic the action of serum to stimulate aromatase expression in these cell types have as yet to be elucidated. Clones from intestine also predominantly contained exon I.4, as did a few from brain. Harada et al., (1993) also reported that a sequence identical to the one we call I.4 was present in transcripts from fetal liver and brain. Sequence-specific Northern and PCR analysis of fetal testes and ovary, and Sertoli cell tumors obtained from patients with Peutz-Jegher Syndrome, a condition characterized by the presence of estrogen-producing bilateral multifocal sex cord tumors, revealed mainly promoter II-specific sequence, similar to the situation in adult ovary (Bulun et al., 1993b). Interestingly, in an estrogen-secreting hepatocellular carcinoma, aromatase expression was driven by promoter II in contrast to the situation in fetal liver, or in adult liver where expression is undetectable (Bulun, unpublished observations). No exon I.1-specific sequences (the predominant sequence in placenta) were present in any clones isolated from these libraries.

A major finding in the last few years was the discovery of a new exonic sequence expressed in the brain of rat, monkey, and human (Honda et al., 1994; Mouri et al., 1995). This "brain-specific" sequence is the major 5'-terminus of transcripts in rat amygdala, and is also present in transcripts in the hypothalamus-preoptic area (HPOA) (Kato et al., 1996). Regulation of aromatase in brain differs from that in other tissues in that expression appears to be increased by androgens and either suppressed or not affected by cAMP (Lephart et al., 1992). In cultured cells derived from mouse embryonic hypothalamus, aromatase expression is elevated by α1-adrenergic agonists, but not those selective for α2- or β-adrenergic receptors. Substance P, cholecystokinin, neurotensin, and brain natriuretic peptide as well as phorbol esters and Bt$_2$ cGMP all increased aromatase expression, suggesting a major role of PKC and PKG pathways in this regulation, which is presumably mediated via the brain-specific promoter (Abe-Dohmae et al., 1996).

However, promoter II-specific transcripts have also been detected in amygdala and HPOA regions (Kato et al., 1996) and, as mentioned above, I.4-specific transcripts have also been detected in brain (Toda et al., 1994). So it may be that different promoters are employed in the various brain loci of expression, and that consequently the regulation is quite different in different brain sites. It should also be noted that transcripts derived from the brain-specific promoter have been detected by RT-PCR in other non-neural cells, namely ovary, placenta, and in a human acute monocytic leukemia cell line (THP-1)(Shozu et al., 1996), although these were present in low abundance in these sites.

PHYLOGENY AND ONTOGENY OF AROMATASE EXPRESSION

Based on comparison of the sequences of the P450arom cDNAs with those of other members of the cytochrome P450 superfamily, it appears that P450arom is only distantly related to other steroidogenic forms of P450 and indeed is one of the most ancient of the cytochrome P450 lineages, apparently evolving more than 1,000 million years ago (Nelson et al., 1993). Certainly aromatase is present throughout the entire vertebrate phylum, but to our knowledge, has not been described in invertebrates. It would be of considerable interest therefore to know if the ancestral gene product is expressed in non-vertebrate phyla, and what reaction it catalyzes.

Inactivation of the SF-1 gene in mice by means of homologous recombination has indicated that this transcription factor is a critical developmental factor for the gonads as well as the adrenals, since these organs fail to develop in such animals (Luo et al., 1994). In the normal embryonic mouse, SF-1 expression is first detected in the genital ridge around embryonic day nine (Ikeda et al., 1994). It subsequently is expressed in the developing adrenal cortex. In genotypic males, testicular development is initiated following a burst of sex-determining region on Y (SRY) expression on embryonic day 11.5, and subsequently the testes express SF-1 in both the Leydig cells and in the Sertoli cells. In females, SF-1 is subsequently expressed in the developing ovary. This developmental expression of SF-1 in the embryonic mouse correlates well with the expression of steroidogenic enzymes as well as of anti-Müllerian hormone in the case of the male. In the mouse embryo, SF-1 is also expressed in the cells which give rise to the hypothalamus around embryonic day 11. As the hypothalamus develops, SF-1 is expressed in the ventral medial region. Subsequent to this, aromatase expression in the hypothalamus is detectable and increases with embryonic development reaching a maximum just prior to the onset of gestation (Ikeda et al., 1995).

An interesting question then arises as to the role of SF-1 in regulating the developmental expression of aromatase in the brain. The brain-specific promoter of the human aromatase gene contains an SF-1-like sequence in the intron immediately downstream from the brain-specific 5'-untranslated exon (Honda et al., 1994b). It remains to be determined if this element plays an important role in the expression of aromatase in this region of the brain. Aromatase is also expressed in other regions of the brain including the amygdala and preoptic nucleus. It is not clear whether the brain-specific promoter is responsible for expression of aromatase in all of these regions of the brain, since promoter II- and I.4-specific transcripts have also been detected in various brain sites (Toda et al., 1994; Kato et al., 1996).

Although little is known about expression of the aromatase gene in lower vertebrates, the sequence of an aromatase gene from the rainbow trout was recently published (Tanaka et al., 1995). In this case, the sequence immediately upstream from the start of transcription contained two sites with close homology to the mammalian SF-1 site, probably indicating that this is the promoter region for aromatase in the gonads of fish. Thus, not surprisingly, this proximal promoter is a primordial promoter of aromatase throughout the entire vertebrate phylum. In lower vertebrates such as fish, aromatase is also expressed to a very high level in the brain (Callard et al., 1978, 1980; Callard, 1981). Recently it has been shown that in the goldfish, two transcripts for aromatase exist, one in ovary and one in brain, that differ significantly throughout the entire coding region, indicative of the presence of two separate genes (Callard and Tchoudakova, 1996). Thus the question of the evolution of the human gene with a single coding region, but with multiple untranslated first exons, becomes a very interesting and complex issue. Another variant on this theme occurs in the pig (Corbin et al., 1995) in which placental transcripts differ from those in ovary at several regions in the coding region. This is consistent with one gene with alternatively-spliced coding exons. Thus the phylogenetic evolution of the aromatase gene will likely turn out to be very complex.

It is clear that in most vertebrates aromatase expression is confined to the brain and the gonads. However as previously indicated, in a number of mammals including primates and some ungulate species such as cow, pig, and horse, expression also occurs in the placenta. In all of these species it appears that a unique placental-specific distal promoter is employed (Hinshelwood et al., 1995). However, there appears to be little sequence homology between the placental promoter regions of the human and bovine genes, making it difficult to propose that these arose from a common ancestral sequence. Furthermore, the creation of transgenic mice expressing a reporter gene downstream of the human promoter I.1 sequence has shown that expression of the reporter occurs in the mouse placenta, and only at that site (Graves et al., 1996). Since mouse placenta does not express endogenous aromatase activity this means that, nevertheless, the requisite regulatory proteins and transcription factors are present in the mouse placenta.

Since SF-1 is apparently not expressed in placenta this may explain why the proximal promoter of aromatase is not employed in this fetal tissue but rather a distal promoter is used which is regulated by mechanisms totally independent from those involving SF-1. Thus, whereas SF1 may play a critical role in both the phylogenetic and ontogenetic regulation of aromatase expression, the diversity of promoters of aromatase found in a number of mammalian species may reflect, in part, the need for alternative means of regulating aromatase gene expression in tissues where SF-1 is not present or else is not functional.

ACKNOWLEDGEMENTS

This work was supported by USPHS grants #R37-AG08174 and HD13234 as well as by grant #3660-046 from the Texas Higher Education Coordinating Board Advanced Research Program. S.E.B. was the recipient of an AAOGF Fellowship and M.D.M. was supported in part by USPHS Training Grant #5-T32-HD07190. The authors gratefully acknowledge the skilled editorial assistance of Susan Hepner.

REFERENCES

Abe-Dohmae, S., Takagi, Y., & Harada, N. (1996). Control of Aromatase mRNA in Cultured Diencephalic Neurons,. P. 18. Proc. IV Int. Arom. Conf., Tahoe City, CA.

Ackerman, G. E., Smith, M. E., Mendelson, C. R., MacDonald, P. C., & Simpson, E. R. (1981). Aromatization of androstenedione by human adipose tissue stromal cells in monolayer culture. J. Clin. Endocrinol. Metab. 53, 412-417.

Agarwal, V. R., Ashanullah, C. I., Simpson, E. R., & Bulun, S. E. (1996). Alternatively spliced transcripts of the aromatase cytochrome P450 (*CYP19*) gene in adipose tissue of women. J. Clin. Endo. Metab. 82, 70-74.

Agarwal, V. R., Bulun, S. E., Leitch, M., Rohrich, R., & Simpson, E. R. (1996). Use of alternative promoters to express the aromatase cytochrome P450 (*CYP19*) gene in breast adipose tissues of cancer-free and breast cancer patients. J. Clin. Endo. Metab. 81, 3843-3849.

Berkovitz, G. D., Bisat, T., & Carter, K. M. (1989). Aromatase activity in microsomal preparations of human genital skin fibroblasts: Influence of glucocorticoids. J. Steroid Biochem. 33, 341-347.

Boulton, T. G., Stahl, N. S., & Yancopoulos, G. D. (1994). Ciliary neurotrophic factor/leukemia inhibitory factor/interleukin 6/oncostatin M family of cytokines induces tyrosine phosphorylation of a common set of proteins overlapping those induced by other cytokines and growth factors. J. Biol. Chem. 269, 11648-11655.

Bulun, S. E., Price, T. M., Mahendroo, M. S., Aitken, J., & Simpson, E. R. (1993a). A link between breast cancer and local estrogen biosynthesis suggested by quantification of breast adipose tissue aromatase cytochrome P450 transcripts using competitive polymerase chain reaction after reverse transcription. J. Clin. Endocrin. Metab. 77, 1622-1628.

Bulun, S. E., Rosenthal, I. M., Brodie, A. M. H., Inkster, S. E., Zeller, W. P., DiGeorge, A. M., Frasier, S. D., Kilgore, M. W., & Simpson, E. R. (1993b). Use of tissue-specific promoters in the regulation of aromatase cytochrome P450 gene expression in human testicular and ovarian sex cord tumors, as well as in normal fetal and adult gonads. J. Clin. Endocrinol. Metab. 77, 1616-1621.

Bulun, S. E., & Simpson, E. R. (1994). Competitive RT-PCR analysis indicates levels of aromatase cytochrome P450 transcripts in adipose tissue of buttocks, thighs, and abdomen of women increase with advancing age. J. Clin. Endocrinol. Metab. 78, 428-432.

Callard, G. V., Petro, Z., & Ryan, K. J. (1978). Phylogenetic distribution of aromatase and other androgen-converting enyzmes in the central nervous sytem. Endocrinology 103, 2283-2290.

Callard, G. V., Petro, Z., & Ryan, K. J. (1980). Aromatization and 5a-reduction in brain and non-neural tissues of a cyclostome (Petromyzan marinus). Gen. Compar. Endocrinol. 42, 155-159.

Callard, G. V. (1981). Aromatization is cyclic AMP-dependent in cultured reptilian brain cells. Brain Res. 204, 451-454.

Callard, G. V., Pudney, J. A., Kendell, S. L., & Reinboth, R. (1984). *In vitro* conversion of androgen to estrogen in amphioxus gonadal tissue. Gen. Compara. Endocrinol. 56, 53-58.

Callard, G. V., & Tchoudakova, A. (1996). Evolutionary and Functional Significance of Two CYP19 Gene Differentially Expressed in Brain and Ovary of Goldfish,.p. 20. Proc. IV Int. Arom. Conf., Tahoe City, CA.

Corbin, C. J., Khalil, M. W., & Conley, A. J. (1995). Functional ovarian and placental isoforms of porcine aromatase. Mol. Cell. Endocrinol. 113, 29-37.

Darnell, J. E.,Jr., Kerr, I. M., & Stark, G. R. (1994). Jak-STAT pathways and transcriptional activation in response to IFNs and other extracellular signaling proteins. Science 264, 1415-1420.

Daynes, R. A., Araneo, B. A., Ershler, W. B., Maloney, C., Li, G. Z., & Ryu, S. Y. (1993). Altered regulation of IL-6 production with normal aging. J. Immunol. 150, 5219-5230.

Dickson, R. B., & Lippman, M. E. (1987). Estrogenic regulation of growth and polypeptide growth factor secretion in human breast carcinoma. Endocrine Rev. 8, 29-43.

Edman, C. D., & MacDonald, P. C. (1976). The role of extraglandular estrogen in women in health and disease. In The Endocrine Function of the Human Ovary (James, V.H.T., Serio, M., & Giusti, G., eds.), pp. 135-140. Academic Press, London.

Fitzpatrick, S. L., & Richards, J. S. (1994). Identification of a cyclic adenosine 3',5'-monophosphate-response element in the rat aromatase promoter that is required for transcriptional activation in rat granulosa cells and R2C leydig cells. Mol. Endocrinol. 8, 1309-1319.

France, J. T., & Liggins, G. C. (1969). Placental sulfatase deficiency. J. Clin. Endocrinol. Metab. 29, 138-144.

Graves, K. A., Kunczt, C., Smith, M. E., Simpson, E. R., & Mendelson, C. R. (1996). Identification of Regions of the Human Aromatase P450 Gene Involved in Placenta- and Gonad-Specific Expression Using Transgenic Mice, p.30. Proc.IV Int.Arom.Conf., Tahoe City, CA.

Harada, N., Yamada, K., Saito, K., Kibe, N., Dohmae, S., & Takagi, Y. (1990). Structural characterization of the human estrogen synthetase (aromatase) gene. Biochem. Biophys. Res. Commun. 166, 365-372.

Harada, N. (1992). A unique aromatase (P450arom) mRNA formed by alternative use of tissue specific exons I in human skin fibroblasts. Biochem. Biophys. Res. Comm. 189, 1001-1007.

Harada, N., Utsume, T., & Takagi, Y. (1993). Tissue-Specific Expression of the Human Aromatase Ccytochrome P450 Gene by Alternative Use of Multiple Exons 1 and Promoters, and Switching of Tissue-Specific Exons 1 in Carcinogenesis. Proc. Natl. Acad. Sci. USA 90, 11312-11316.

Hemsell, D. L., Grodin, J. M., Brenner, P. F., Siiteri, P. K., & MacDonald, P. C. (1974). Plasma precursors of estrogen. II. Correlation of the extent of conversion of plasma androstenedione to estrone with age. J. Clin. Endocrinol. Metab. 38, 476-479.

Hilton, D. J., Hilton, A. A., Raicevic, A., Rakar, S., Harrison-Smith, M., Gough, N. M., Begley, C. G., Metcalf, D., Nicola, N. A., & Willson, T. A. (1994). Cloning of a murine IL-11 receptor α-chain; requirement for gp130 for high affinity binding and signal transduction. EMBO J. 13, 4765-4775.

Hinshelwood, M. M., Liu, Z., Conley, A. J., & Simpson, E. R. (1995). Demonstration of tissue-specific promoters in nonprimate species that express aromatase P450 in placentae. Biol. Reprod. 53, 1151-1159.

Honda, S.-I., Harada, N., & Takagi, Y. (1994). Novel exon I of the aromatase gene specific for aromatase transcripts in brain. Biochem. Biophys. Res. Commun. 198, 1153-1160.

Hutchinson, J. B. (1991). Hormonal control of behavior: Steroid action in the brain. Curr. Opin. Neurobiol. 1, 562-570.

Ikeda, Y., Shen, W., Ingraham, H. A., & Parker, K. L. (1994). Developmental expression of mouse steroidogenic factor-1, an essential regulator of the steroid hydroxylases. Mole. Endocrinol. 9, 656-662.

Ikeda, Y., Luo, X., Abbud, R., Nilson, J. H., & Parker, K. L. (1995). The nuclear receptor steroidogenic factor-1 is essential for the formation of the ventromedial hypothalamic nucleus. Mol. Endocrinol. 9, 478-486.

Jenkins, C., Michael, D., Mahendroo, M., & Simpson, E. (1993). Exon-specific northern analysis and rapid amplification of cDNA ends (RACE) reveal that the proximal promoter II (PII) is responsible for aromatase cytochrome P450 (CYP19) expression in human ovary. Mol. Cell. Endocrinol. 97, R1-R6.

Kato, J., Mouri, Y. N., & Hirata, S. (1996). Analysis of Structure of Aromatase in RNA in the Rat Brain, p. 20. Proc. IV Int. Arom. Conf., Tahoe City, CA.

Keller, D. C., Du, X. X., Srour, E. F., Hoffman, R., & Williams, D. A. (1993). Interleukin-11 inhibits adipogenesis and stimulates myelopoiesis in human long-term marrow cultures. Blood 82, 1428-1435.

Kellis, J. T.,Jr., & Vickery, L. E. (1987). Purification and characterization of human placental aromatase cytochrome P450. J. Biol. Chem. 262, 4413-4420.

Kilgore, M. W., Means, G. D., Mendelson, C. R., & Simpson, E. R. (1992). Alternative promotion of aromatase cytochrome P450 expression in human fetal tissues. Mol. Cell. Endocrinol. 83, R9-R16.

Killinger, D. W., Perel, E., Daniilescu, D., Kherlip, L., & Lindsay, W. R. N. (1987). The relationship between aromatase activity and body fat distribution. Steroids 50, 61-72.

Kishimoto, T., Akira, S., & Taga, T. (1992). Interleukin-6 and its receptor: a paradigm for cytokines. Science 258, 593-597.

Lala, D. S., Rice, D. A., & Parker, K. L. (1992). Steroidogenic factor 1, a key regulation of steroidogenic gene expression, is the mouse homolog of fushi tarazu-factor 1. Mol. Endocrinol. 6, 1249-1258.

Lanoux, M. J., Cleland, W. H., Mendelson, C. R., Carr, B. R., & Simpson, E. R. (1985). Factors affecting the conversion of androstenedione to estrogens by human fetal hepatocytes in monolayer culture. Endocrinology 117, 361-367.

Lephart, E. D., Simpson, E. R., & Ojeda, S. R. (1992). Effects of cyclic AMP and androgens on *in vitro* brain aromatase enzyme activity during prenatal development in the rat. Journal of Neuroendocrinol. 4, 29-35.

Luo, X., Ikeda, Y., & Parker, K. L. (1994). A cell-specific nuclear receptor is essential for adrenal and gonadal development and sexual differentiation. Cell 77, 481-490.

Mahendroo, M. S., Means, G. D., Mendelson, C. R., & Simpson, E. R. (1991). Tissue-specific expression of human P450$_{arom}$: the promoter responsible for expression in adipose is different from that utilized in placenta. J. Biol. Chem. 266, 11276-11281.

Mahendroo, M. S., Mendelson, C. R., & Simpson, E. R. (1993). Tissue-specific and hormonally-controlled alternative promoters regulate aromatase cytochrome P450 gene expression in human adipose tissue. J. Biol. Chem. 268, 19463-19470.

Means, G. D., Mahendroo, M., Corbin, C. J., Mathis, J. M., Powell, F. E., Mendelson, C. R., & Simpson, E. R. (1989). Structural analysis of the gene encoding human aromatase cytochrome P-450, the enzyme responsible for estrogen biosynthesis. J. Biol. Chem. 264, 19385-19391.

Means, G. D., Kilgore, M. W., Mahendroo, M. S., Mendelson, C. R., & Simpson, E. R. (1991). Tissue-specific promoters regulate aromatase cytochrome P450 gene expression in human ovary and fetal tissues. Mol. Endocrinol. 5, 2005-2013.

Mendelson, C. R., Wright, E. E., Porter, J. C., Evans, C. T., & Simpson, E. R. (1985). Preparation and characterization of polyclonal and monoclonal antibodies against human aromatase cytochrome P-450 (P-450$_{arom}$), and their use in its purification. Arch. Biochem. Biophys. 243, 480-491.

Michael, M. D., Kilgore, M. W., Morohashi, K.-I., & Simpson, E. R. (1994). Ad4BP/SF-1 regulates cyclic AMP-induced transcription from the proximal promoter (PII) of the human aromatase P450 (CYP19) gene in the ovary. J. Biol. Chem. 270, 13561-13466.

Michael, M. D., & Simpson, E. R. (1996). A CRE-Like Sequence in Promoter II is Necessary for Cyclic AMP-Induced Transcription of the Human Aromatase (CYP19) Gene in the Ovary. Proc. of the 10th Int. Cong. Endocrinol. p.96. San Francisco, CA.

Miller, W. R., & O'Neill, J. (1987). The importance of local synthesis of estrogen within the breast. Steroids 50, 537-548.

Mouri, Y. N., Hirata, S., & Kato, J. (1995). Analysis of the expression and the first exon of aromatase mRNA in monkey brain. J. Steroid Biochem. Mol. Biol. 55, 17-23.

Nakajin, S., Shimoda, M., & Hall, P. F. (1986). Purification to homogeneity of aromatase from human placenta. Biochem. Biophys. Res. Commun. 134, 704-710.

Narazaki, M., Witthahn, B. A., Yoshida, K., Silvennoinen, O., Yasukawa, K., Ihle, J. N., Kishimoto, T., & Taga, T. (1994). Activation of JAK2 Kinase Mediated by the Interleukin-6 Signal Transducer gp130. Proc. Natl. Acad. Sci. USA 91, 2285-2289.

Nelson, D. R., Kamataki, T., Waxman, D. J., Guengerich, F. P., Estabrook, R. W., Feyereisen, R., Gonzalez, F. J., Coon, M. J., Gunsalus, I. C., Gotoh, O., Okuda, K., & Nebert, D. W. (1993). The P450 superfamily: Update on new sequences, gene mapping, accession numbers, early trivial names of enzymes, and nomenclature. DNA Cell Biol. 12, 1-51.

O'Neill, J. S., Elton, R. A., & Miller, W. R. (1988). Aromatase activity in adipose tissue from breast quadrants: a link with tumor site. British Med. J. 296, 741-743.

Price, T., Aitken, J., Head, J., Mahendroo, M. S., Means, G. D., & Simpson, E. R. (1992a). Determination of aromatase cytochrome P450 messenger RNA in human breast tissues by competitive polymerase chain reaction (PCR) amplification. J. Clin. Endocrinol. Metab. 74, 1247-1252.

Price, T., O'Brien, S., Dunaif, A., & Simpson, E. R. (1992b). Comparison of aromatase cytochrome P450 mRNA levels in adipose tissue from the abdomen and buttock using competitive polymerase chain reaction amplification. Proc. Soc. Gynecol. Invest. 179 (Abstract).

Reed, M. J., Topping, L., Coldham, N. G., Purohit, A., Ghilchik, M. W., & James, V. H. T. (1993). Control of aromatase activity in breast cancer cells: the role of cytokines and growth factors. J. Steroid Biochem. Mol. Biol. 44, 589-596.

Sasano, H., Nagura, H., Harada, N., Goukon, Y., & Kimura, M. (1994). Immunolocalization of aromatase and other steroidogenic enzymes in human breast disorders. Human Pathol. 25, 530-535.

Schindler, C., Fu, X.-Y., Improta, T., Aebersold, R., & Darnell, J. E., Jr. (1992). Proteins of Transcription Factor ISGF-3: One Gene Encodes the 91-and 84 kDa

ISGF-3 Proteins that are Activated by Interferon Alpha. Proc. Ntl. Acad. Sci. USA 89, 7836-7839.

Schrey, M. P., & Patel, K. V. (1995). Prostaglandin E_2 production and metabolism in human breast cancer cells and breast fibroblasts. Regulation by inflammatory mediators. British J. Cancer 72, 1412-1419.

Shozu, M., Zhao, Y., & Simpson, E. R. (1996). Regulation of Aromatase Activity in the THP-1 Osteoclastic Cell Line, Proc. of the 10th International Congress Endocrinology, p. 59. San Franscisco.

Simpson, E. R., Ackerman, G. E., Smith, M. E., & Mendelson, C. R. (1981). Estrogen Formation in Stromal Cells of Adipose Tissue of Women: Induction by Glucocorticosteroids. Proc. Natl. Acad. Sci. USA 78, 5690-5694.

Simpson, E. R., Mahendroo, M. S., Means, G. D., Kilgore, M. W., Hinshelwood, M. M., Graham-Lorence, S., Amarneh, B., Ito, Y., Fisher, C. R., Michael, M. D., Mendelson, C. R., & Bulun, S. E. (1994). Aromatase cytochrome P450, the enzyme responsible for estrogen biosynthesis. Endocrine Rev. 15, 342-355.

Stahl, N., & Yancopoulos, G. D. (1993). The alphas, betas and kinases of cytokine receptor complexes. Cell 74, 587-590.

Stahl, N., Boulton, T. G., Farruggella, T., Ip, N. Y., Davis, S., Witthuhn, B. A., Quelle, F. W., Silvennoinen, O., Barbieri, G., Pellegrini, S., Ihle, J. N., & Yancopoulos, G. D. (1994). Association and activation of JAK-TYK kinases by CNTF-LIF-OSM-IL-6β receptor components. Science 263, 92-95.

Stahl, N., Farruggella, T., Boulton, T. G., Zhong, Z., Darnell, J. E., & Yancopoulos, G. D. (1995). Modular tyrosine-based motifs in cytokine receptor specify choice of STATs and other substrates. Science 267, 1349-1353.

Stephanon, A., Sarlis, N. J., Richards, R., & Handwerger, S. (1994). Expression of retinoic acid receptor subtypes and cellular retinoic acid binding protein II in RNAs during differentiation of human trophoblast cells. Biochem. Biophys. Res. Commun. 202, 772-780.

Sun, T., Zhao, Y., Fisher, C. R., Kilgore, M., Mendelson, C. R., & Simpson, E. R. (1996). Characterization of *cis*-Acting Elements of the Human Aromatase P450 (CYP19) Gene that Mediate Regulation by Retinoids in Human Choriocarcinoma Cells. Proc. of the 10th Int. Cong. Endocrinol., p. 118. San Francisco.

Tanaka, M., Fukada, S., Matsuyama, M., & Nagahama, Y. (1995). Structure and promoter analysis of the cytochrome P450 aromatase gene of the teleost fish, medaka (Oryzias latipes). J. Biochem. 117, 719-725.

Thompson, E. A.,Jr., & Siiteri, P. K. (1974). The involvement of human placental microsomal cytochrome P450 in aromatization. J. Biol. Chem. 249, 5373-5378.

Toda, K., Terashima, M., Kamamoto, T., Sumimoto, H., Yamamoto, Y., Sagara, Y., Ikeda, H., & Shizuta, Y. (1990). Structural and functional characterization of human aromatase P450 gene. Eur. J. Biochem. 193, 559-565.

Toda, K., Miyahara, K., Kawamoto, T., Ikeda, H., Sagara, Y., & Shizuta, Y. (1992). Characterization of a cis-acting regulatory element involved in human aromatase P450 gene expression. Euro. J. Biochem. 205, 303-309.

Toda, K., & Shizuta, Y. (1993). Molecular cloning of a cDNA showing alternative splicing of the 5'-untranslated sequence of mRNA for human aromatase P450. Eur. J. Biochem. 213, 383-389.

Toda, K., Simpson, E. R., Mendelson, C. R., Shizuta, Y., & Kilgore, M. W. (1994). Expression of the gene encoding aromatase cytochrome P450 (CYP19) in fetal tissues. Mol. Endocrinol. 8, 210-217.

Wei, J., Xu, H., Davies, J. L., & Hemmings, G. P. (1992). Increase of plasma IL-6 concentration with age in healthy subjects. Life Sci. 51, 1953-1956.

Yamada, K., Harada, N., Honda, S., & Takagi, Y. (1995). Regulation of placenta-specific expression of the aromatase cytochrome P450 gene. Involvement of the trophoblast-specific element binding protein. J. Biol. Chem. 270, 25064-25069.

Zhao, Y., Mendelson, C. R., & Simpson, E. R. (1995a). Characterization of the sequences of the human CYP19 (aromatase) gene that mediate regulation by glucocorticoids in adipose stromal cells and fetal hepatocytes. Mol. Endocrinol. 9, 340-349.

Zhao, Y., Nichols, J. E., Bulun, S. E., Mendelson, C. R., & Simpson, E. R. (1995b). Aromatase P450 gene expression in human adipose tissue: Role of a Jak/STAT pathway in regulation of the adipose-specific promoter. J. Biol. Chem. 270, 16449-16457.

Zhao, Y., Agarwal, V. R., Mendelson, C. R., & Simpson, E. R. (1996a). Estrogen biosynthesis proximal to a breast tumor is stimulated by PGE_2 via cyclic AMP, leading to activation of promoter II of the CYP19 (aromatase gene). Endocrinol. 137, 5739-5742.

Zhao, Y., Nichols, J. E., Valdez, R., Mendelson, C. R., & Simpson, E. R. (1996b). TNFa stimulates aromatase gene expression in human adipose stromal cells (ASC) through use of an AP-1 binding site upstream of promoter I.4. Mol. Endocrin. 10, 1350-1357.

Zhong, Z., Wen, Z., & Darnell, J. E. (1994). STAT3 and STAT4: Members of the Family of Signal Transducers and Activators of Transcription. Proc. Natl. Acad. Sci. USA 91, 4806-4810.

Chapter 6

Molecular Aspects of Precocious Puberty

WAI-YEE CHAN and GORDON B. CUTLER, JR.

INTRODUCTION

Puberty is the last phase of the complex process of sexual maturation, a process by which an individual acquires reproductive competency. Puberty is considered unequivocally precocious when any sign of sec-

Advances in Molecular and Cellular Endocrinology
Volume 2, pages 121-141.
Copyright © 1998 by JAI Press Inc.
All rights of reproduction in any form reserved.
ISBN: 0-7623-0292-5

ondary sexual maturation appears at an age more than three standard deviations below the mean, i.e., before the age of eight in a girl or nine in a boy (Cutler, 1992, 1993; Grumbach and Styne, 1992). Normal sexual development at puberty is dependent on the integrity of the hypothalamic-pituitary-gonadal axis. Increased secretion of luteinizing hormone-releasing hormone (LHRH) leads to increased secretion of gonadotropins and sex steroids. In girls, the outward and visible sign of estradiol secretion is breast development. The prime manifestation of gonadotropin secretion in boys is testicular enlargement. Secondary sexual characteristics which do not include breast or testicular enlargement do not constitute precocious puberty and alternative explanation should be considered (Brook, 1995). If the precocious puberty results from premature reactivation of the hypothalamic LHRH pulse generator-pituitary gonadotropin-gonadal axis, the condition is called complete isosexual precocious puberty, or true or central precocious puberty, and is LHRH-dependent. If extrapituitary secretion of gonadotropins or secretion of gonadal steroids independent of pulsatile LHRH stimulation leads to virilization in boys or feminization in girls, the condition is termed incomplete isosexual precocious puberty, or pseudoprecocious puberty, or LHRH-independent precocious puberty (Grumbach and Styne, 1992). In these conditions, the hypothalamic-pituitary-gonadal axis is not active. Increased amounts of circulating sex steroids in patients with precocious puberty lead to increases in height velocity, somatic development, skeletal maturation, and to early fusion of the epiphyses. This results in tall stature during childhood but short stature in adulthood.

LHRH-DEPENDENT PRECOCIOUS PUBERTY

Little is known about the mechanism of LHRH-dependent precocious puberty. There is an early appearance of the mature function of the LHRH neuron. Physiologically, there is a large increase in gonadotropin pulse amplitude (Oerter et al., 1990), and perhaps also an increase in pulse frequency (Kelch et al., 1985; Dunkel et al., 1992). The regulatory mechanism of LHRH release by the LHRH neurons is unknown. Knowledge is somewhat more advanced for the synthesis of LHRH and its receptor. cDNA cloning revealed that LHRH is a decapeptide derived from a precursor of 90 amino acids (Seeburg and Adelman, 1984). The human LHRH receptor cDNA was also cloned (Kakar et al., 1992). It is a member of the G-protein-coupled receptor family with the characteristic seven-transmembrane (TM) helical domain. However, in spite of these studies, no mutation of either the hormone or its receptor has been identified in patients with central precocious puberty. Therefore, currently, we do not understand the molecular etiology of any forms of LHRH-dependent precocious puberty.

LHRH-INDEPENDENT PRECOCIOUS PUBERTY

Noncentral forms of precocious puberty include those mediated by chorionic gonadotropin (hCG)-producing tumors, tumors of the adrenal gland or ovary, virilizing forms of adrenal hyperplasia, the McCune-Albright syndrome (MAS), and familial male-limited precocious puberty. Discussion on the molecular aspects of precocious puberty in this chapter will be limited to that of the latter three conditions with emphasis on MAS and familial male-limited precocious puberty.

Virilizing Congenital Adrenal Hyperplasia

Congenital adrenal hyperplasia is a family of autosomal recessive disorders of adrenal steroidogenesis, the most common form of which is 21-hydroxylase deficiency. In 21-hydroxylase deficiency there is impaired biosynthesis of cortisol and aldosterone and overproduction of cortisol precursors that do not require 21-hydroxylation for their biosynthesis. Some of these precursors are shunted into the androgen pathway and cause the signs and symptoms of androgen excess. Clinically, there are two major forms of 21-hydroxylase deficiency, namely, the classic and the nonclassic (or late-onset) forms. The classic form is further subdivided into the salt-losing variant and the simple virilizing variant. These different clinical forms of 21-hydroxylase deficiency represent a continuous spectrum of disease severity. The distinction between these forms on a molecular basis is not absolute and merely represents phenotypic expression of different mutations of a single active 21-hydroxylase gene (White et al., 1985).

Molecular Genetic Studies

The frequency of genetic abnormalities involving 21-hydroxylase is due to a nearby pseudogene that arises from an ancestral gene duplication event. The active gene (containing 10 exons) to avoid ambiguity and a pseudogene are located in tandem on the short arm of chromosome 6 (White et al., 1985). These two genes are 96% homologous in their introns and 98% homologous in their coding sequences, with very little polymorphism in either of the two genes. Both genes are actively transcribed but the pseudogene lacks an open reading frame and does not encode a protein. Much is known about the molecular genetics of 21-hydroxylase deficiency. Most patients with 21-hydroxylase deficiency do not have homozygous mutations in the active gene, but rather are compound heterozygotes with different mutations in each of the active alleles. A number of mutations in the active 21-hydroxylase gene have been identified. The majority of the mutations are gene conversions in which small deleterious mutations are transferred from the inactive pseudogene to the active gene via genetic events such as unequal cross-over and homologous recombination. Deletions and single base substitutions of the active gene account for the remaining mutations (White and New, 1988; White et al., 1988; White, 1989). Alterations of the gene occur in exons as well as introns leading to genetic events such as frame shift, splicing variation, premature truncation, etc. Advances in our understanding of the molecular genetics of congenital adrenal hyperplasia is summarized in a recent review by Laue and Rennert (1995).

Although molecular studies of MAS and familial male-limited precocious puberty (FMPP) lagged behind those of 21-hydroxylase deficiency, considerable recent progress has been made in elucidating the molecular genetic events leading to these two forms of LHRH-independent precocious puberty.

McCune-Albright Syndrome

MAS is a sporadic disease classically defined by precocious puberty, polyostotic fibrous dysplasia, café-au-lait pigmentation of skin, and multiple endocrinopathies, including hyperthyroidism, hypercortisolism, hyperprolactinemia, and growth hormone excess (Schwindinger and Levine, 1993). The pigmented cutaneous lesions typically display a segmental distribution that frequently follows the lines of Blaschko (Happle, 1986). Because of the sporadic occurrence of MAS, the segmental distribution of the bone lesions, the cutaneous hyperpigmentation, and the variability in metabolic abnormalities, it has been postulated that this disorder is due to a dominant spontaneous postzygotic somatic mutation occurring early in embryogenesis. The mosaic distribution of the mutation-bearing cells accounts for the varied clinical presentation of the symptoms in multiple systems (Happle, 1986).

The metabolic abnormalities in MAS are characterized by hyperactive endocrine glands with autonomous function. The diverse syndromes of glandular hyperfunction have in common the involvement of cells that respond to extracellular signals through the activation of the hormone-sensitive adenylyl cyclase/cAMP signaling pathway. The fact that these metabolic disturbances are not accompanied by elevated plasma concentration of the relevant trophic or stimulatory hormones has led to the speculation that MAS is caused by a lesion in components of the signal transduction pathway that control the production of cAMP. This speculation subsequently led to the discovery of mutations in the stimulatory G [guanine nucleotide-binding] protein G_s, more specifically, the α subunit of G_s which has intrinsic guanosine triphosphatase (GTPase) activity and inactivates the G protein by hydrolyzing the bound GTP to GDP. In MAS, mutations that inhibit the GTPase activity of $G_s\alpha$ prolong the lifetime of GTP-bound α subunit and result in increased activity of the adenylyl cyclase (Levine, 1991). Such somatic activating mutations of $G_s\alpha$ have been identified in human growth hormone-secreting pituitary adenomas and thyroid tumors and the mutations convert the $G_s\alpha$ into a putative oncogene termed *gsp* (Landis et al., 1989). The identification of activating mutations in the $G_s\alpha$ gene in patients with MAS and the documentation of the variable distribution of cells containing these mutations are important milestones in the elucidation of the molecular basis of this disorder.

Molecular Genetic Studies

By analyzing genomic DNA using primers specific of the $G_s\alpha$ gene, Weinstein et al. (1991) identified two mutations within exon 8 of the gene in tissues from four MAS patients, including affected endocrine organs and tissues not classically involved in MAS. In two of the patients, amino acid residue Arg201 is replaced by His due to a G to A transition and in the other two patients Cys is substituted for the same Arg residue due to a C to T transition. Similar mutations of the $G_s\alpha$ in sporadic growth hormone-secreting pituitary adenomas causes a 30-fold decrease in intrinsic GTPase activity. Cell membranes containing these altered proteins produce cAMP at an elevated rate in the absence of any stimulatory hormone (Landis et al., 1989) leading to cellular hyperplasia and adenomas (Dumont et al., 1989). Thus, the authors concluded that the presence of these activated $G_s\alpha$ and increased cAMP formation in various tissues underlie the clinical manifestations of MAS in these patients (Weinstein et al., 1991). Further studies in 12 other MAS patients show that either the Arg201His or the Arg201Cys is present (Schwindinger et al., 1992; Shenker et al., 1993b, 1994, 1995; Malchoff et al., 1994) confirming the potential causative role of these activating mutations of the $G_s\alpha$ in MAS. A recent report provided clinical evidence for altered adenylyl cyclase activity in MAS (Zung et al., 1995). Many patients with MAS have hypophosphatemia, possibly caused by the presence of the activating $G_s\alpha$ mutation in proximal renal tubules or the elaboration of a phosphaturic factor from fibrous dysplasia. When the urinary cAMP in MAS patients was measured, these authors found an elevated basal urinary cAMP level, but a blunted urinary cAMP response to parathyroid hormone stimulation, suggesting the presence of the activating $G_s\alpha$ mutation in the renal tubules.

The $G_s\alpha$ mutations are present in virtually all affected endocrine as well nonendocrine tissues. In spite of the initial failure in identifying the $G_s\alpha$ mutations in café-au-lait spots and dysplastic bone (Weinstein et al., 1991), later studies confirmed the presence of these mutations in affected skin (Schwindinger et al., 1992), dysplastic bone (Malchoff et al., 1994; Shenker et al., 1994), and osteoblastic cells derived from isolated lesions of fibrous dysplasia (Shenker et al., 1995). In all cases reported, specific regions of abnormal tissues contain a higher proportion of cells with mutant alleles than do adjacent normal tissues. Moreover, $G_s\alpha$ mutations are absent in at least one type of tissue from each patient, indicating that these mutations are somatic rather than germ-line. Furthermore, in each patient, only one specific mutation (either Arg201His or Arg201Cys) is detected, a finding that is consistent with the presence of a single monoclonal population of abnormal cells in an individual (Weinstein et al., 1991).

Genotype-Phenotype Correlation

The genotype-phenotype correlation between $G_s\alpha$ mutations and MAS is not yet totally clear. The existence of a severely affected subset of patients with MAS who may be at risk for nonendocrine disease and sudden or premature death (Shenker et al., 1993b) suggests the possible involvement of other factors besides mutated $G_s\alpha$ in causing MAS in this subset of patients. There are also apparently normal tissues in MAS patients in which the $G_s\alpha$ mutations are present (Weinstein et al., 1991). Furthermore, a clinical condition with the combination of precocious puberty and pseudohypoparathyroidism Ia (PHP-Ia) in two unrelated boys was recently reported to be the result of another $G_s\alpha$ mutation, Ala366Ser (Iiri et al., 1994). This mutation constitutively activates adenylyl cyclase at 20 °C *in vitro* , causing hormone-independent cAMP accumulation when expressed in cultured cells and accounting for the testotoxicosis phenotype. The mutated $G_s\alpha$ is quite stable at the lower temperatures but is rapidly degraded at 37 °C . This is thought to explain the constitutive activity of $G_s\alpha$ in the testes (due to their lower temperature), causing precocious puberty and decreased activity of $G_s\alpha$ at 37 °C , causing resistance to parathyroid hormone. Thus, a single mutation of the $G_s\alpha$ gives rise to both constitutive activity and thermolability of the protein resulting in both gain and loss of endocrine function.

Familial Male-Limited Precocious Puberty

FMPP is a form of isosexual precocious puberty in boys, whose mechanism is independent of the normal pubertal process involving LHRH (Schedewie et al., 1981; Cutler, 1993). An autosomal dominant mutation leads to progressive testicular activation, usually at two to three years of age, without activation of either the ovaries or adrenal glands. Signs of puberty usually appear by 3–4 years of age. Histologic examination of testicular biopsy shows hyperplasia of Leydig cells, hence the term testotoxicosis, indicating primary hyperplasia of the Leydig cells independent of classic gonadotropin stimulation as the cause of this disorder (Rosenthal et al., 1983). Affected boys have secondary sexual development with penile growth and bilateral enlargement of testes and pubic hair development often indistinguishable from true precocious puberty on the basis of physical examination alone (Grumbach and Styne, 1992). Diagnosis is made by the findings of pubertal to normal adult levels of testosterone in conjunction with prepubertal levels of basal and LHRH-stimulated gonadotropins with normal clearance of testosterone. There is the lack of a pubertal pattern of LH pulsatility, whether measured by immunological or bioassay techniques. The increased testosterone production is of testicular origin and not adrenal. Treatment with an LHRH agonist does not suppress the testicular function or maturation (Holland, 1991; Grumbach and Styne, 1992). In adulthood, affected individuals achieve a normal adult pattern of LH secretion and normal gonadotropin response to LHRH. Fertility is the rule (Grumbach and Styne, 1992; Kletter and Kelch, 1993). The pattern of inheritance is autosomal dominant, although sporadic cases occur. In such cases, it is referred to as sporadic male-limited precocious puberty (SMPP). No endocrine abnormalities have been reported of obligate carrier females.

Molecular Genetic Studies

In the testes, human luteinizing hormone/chorionic gonadotropin receptor (hLHR) mediates the effects of LH on testosterone biosynthesis by stimulating the Leydig cell Gs-adenylyl cyclase system. It was hypothesized that a mutation leading to constitutive activation of the hLHR-mediated signaling pathway would explain the pathophysiology of FMPP. The mutant hLHR is also present on ovarian cells in obligate carrier females. The observation that female carriers of these receptor mutations do not manifest clinical signs of precocious puberty is easily explained since both LH and follicle-stimulating hormone (FSH) are required for ovarian steroidogenesis. Hence, constitutive activation of the hLHR would not be expected to result in precocious puberty in females. This is consistent with observations that hCG-secreting tumors (pinealomas, hepatoblastomas) cause sexual precocity only in males and not in females (Sklar et al., 1981).

The first report identifying the molecular defect underlying FMPP was published in 1993 (Shenker et al., 1993a). A missense mutation that results in substitution of Gly for Asp578 in the transmembrane (TM) VI domain of the hLHR was found in affected members of eight different families. Subsequent studies have revealed a total of 11 mutations within exon 11 of the hLHR gene among 64 kindreds of FMPP/SMPP (Kremer et al., 1993; Boepple et al., 1994; Yano et al., 1994, 1995, 1996; Kawate et al., 1995; Kosugi et al., 1995; Kraaij et al., 1995; Latronico et al., 1995; Laue et al., 1995, 1996; Müller et al., 1995; Cocco et al., 1996; Evans et al., 1996; Wu and Leschek, unpublished observations in two families) (Figure 1); seven different mutations found in the 15 cases of SMPP (Boepple et al., 1994; Yano et al., 1994; Laue et al., 1995, 1996; Müller et al., 1995; Evans et al., 1996; Wu and Leschek, unpublished observation in one family); and seven mutations in the 48 families with FMPP (Kremer et al., 1993; Shenker et al., 1993a; Kawate et al., 1995; Kosugi et al., 1995; Kraaij et al., 1995; Laue et al., 1995, 1996; Yano et al., 1995, 1996; Cocco et al., 1996; Evans et al., 1996; Wu and Leschek, unpublished observation in one family). The inheritance of one case is not known (Latronico et al., 1995). The nucleotide substitution, amino acids af-

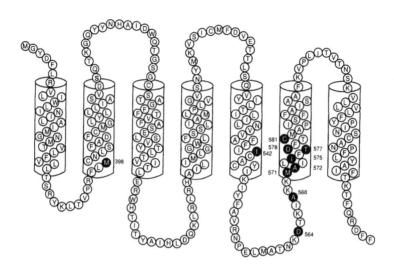

Figure 1. Location of mutated amino acids identified in FMPP patients. The seven TM helices are shown. Filled circles represent amino acid residues found to be mutated in FMPP/SMPP patients. Location of the amino acid residues are as described by Minegishi et al., 1990.

fected, location of the mutations, and the genetics of the different mutations are presented in Table 1. The most frequent mutation observed is Asp578Gly, representing about 67% of all cases of FMPP (sporadic cases included). Four of the mutations, namely those leading to Met398Thr, Ile542Leu, Asp578Gly, and Ala572Val, have been found in both sporadic and familial cases of the disorder. Seven out of the 15 cases of SMPP (47%) are due to the predominant Asp578Gly mutation while about 75% of cases of FMPP result from this mutation. Thus, compared to FMPP, there is greater genetic heterogeneity of activating mutations of the hLHR gene in SMPP. Nonetheless, there does not appear to be a fundamental difference between the sporadic and familial disorder, indicating that the sporadic cases might represent *de novo* germ-line mutations that will give rise to future familial cases. Of the 11 known mutations, six (Met398Thr, Ala568Val, Ala572Val, Ile575Leu, Asp578Gly, and Cys581Arg) were found among seven patients of Asian, Brazilian, or African-American descent (Yano et al., 1994, 1995, 1996; Latronico et al., 1995; Laue et al., 1995, 1996), while the remaining mutations were identified among Caucasian patients from Europe and America. Thus,

Table 1.   Mutations of the hLHR Identified in FMPP Patients

Nucleotide Change	Amino Acid Change	Location	Number of Cases	Genetics	
T 1193 C	Met398Thr	TM II	4	Familial	(3)
				Sporadic	(1)
A 1624 C	Ile542Leu	TM V	4	Familial	(3)
				Sporadic	(1)
A 1691 G	Asp564Gly	3rd cytoplasmic loop	1	Sporadic	
C 1703 T	Ala568Val	3rd cytoplasmic loop	1	?	
G 1713 A	Met571Ile	TM VI	2	Familial	
C 1715 T	Ala572Val	TM VI	2	Familial	(1)
				Sporadic	(1)
A 1723 C	Ile575Leu	TM VI	1	Sporadic	
C 1730 T	Thr577Ile	TM VI	2	Familial	
G 1732 T	Asp578Tyr	TM VI	3	Sporadic	
A 1733 G	Asp578Gly	TM VI	43	Familial	(36)
				Sporadic	(7)
T 1741 C	Cys581Arg	TM VI	1	Familial	

Note: The nucleotide and amino acid positions are as described by Minegishi et al., 1990. TM, transmembrane helices. ? indicates that inheritance of FMPP in the patient is not reported.

patients from non-Caucasian ethnic backgrounds appear to have a relatively high likelihood of having new mutations. Furthermore, there are patients with FMPP in whom mutation of the hLHR has not been identified (Kremer et al., 1993; Boepple et al., 1994; Kosugi et al., 1995; Laue et al., 1996), suggesting that mutation of the hLHR gene might not be the only genetic event leading to such disease.

The increase in plasma testosterone levels in response to exogenous hCG stimulation is similar in normal boys and in boys with FMPP (Oerter et al., 1990). Therefore a constitutively activated hLHR would be expected to respond to high concentrations of ligand and produce a similar response as the wild-type receptor. Expression of all the mutant hLHRs in either COS-7 cells or HEK 293 cells result in increased intracellular levels of cAMP in the absence of agonist, suggestive of constitutive activation of the mutated hLHRs. Hormone stimulation studies reveal that the majority of the mutated hLHRs are hormone-responsive with the exception of the hLHRs carrying the mutations Ile542Leu and Cys581Arg (Shenker et al., 1993a; Yano et al., 1994, 1995, 1996; Kosugi et al., 1995; Kraaij et al., 1995; Latronico et al., 1995; Laue et al., 1995, 1996; Müller et al., 1995). Basal cAMP levels of transfected cells expressing these two mutant hLHRs are higher than those expressing wild-type hLHR and are unaffected by the presence of increasing concentrations of hCG in the culture media (Laue et al., 1995). Whether this difference in ligand responsiveness is caused by the uncoupling of the LHR due to these two mutations or other unknown mechanism is not clear at the present time. The maximum levels of intracellular cAMP achieved in the transfected cells are either comparable to (Shenker et al., 1993a; Yano et al., 1994, 1995, 1996; Kosugi et al., 1995) or lower than (Kosugi et al., 1995; Kraaij et al., 1995; Latronico et al., 1995; Laue et al., 1995, 1996) the levels in cells expressing the wild-type receptor. Figure 2 shows several examples of basal and maximally stimulated intracellular cAMP levels in HEK 293 cells transfected with mutant hLHR cDNA constructs. The EC_{50} in all cases are quite comparable between the wild-type and mutant hLHR transfectants. Similarly, both the agonist affinity and the concentration of hCG-binding sites in cells expressing the mutant hLHR carrying the Ala568Val and the Asp578Gly mutations are unaltered (Yano et al., 1994; Latronico et al., 1995). On the other hand, the Met398Thr mutation confers a higher agonist affinity to the hLHR in spite of having a comparable concentration of hCG-binding sites between transfected cells expressing this mutated receptor and the wild-type receptor (Yano et al., 1995, 1996). However, in a number of cases, even though the agonist affinity of the hLHR is unaffected by the mutation (Met571Ile, Ala572Val, Ile575Leu, Thr577Ile), the concentration of hCG-binding sites present on the cells expressing these mutant hLHRs are significantly reduced when compared to those expressing the wild-type hLHR (Kosugi et al., 1995; Laue et al., 1996).

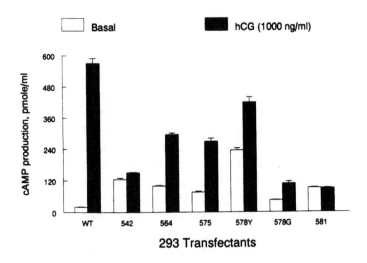

Figure 2. Basal and maximal hCG-stimulated cAMP levels in HEK 293 cells transfected with wild-type or mutant hLHR cDNA constructs. Basal and hCG-stimulated cAMP production are normalized for transformation efficiency. Means ±SEM from triplicate observations are shown (Laue et al., 1995, 1996). WT, Wild-type cDNA; 542, Ile542Leu; 564, Asp564Gly; 575, Ile575Leu; 578Y, Asp578Tyr; 578G, Asp578Gly; 581, Cys581Arg.

The activity of the mutant hLHR in the phosphatidylinositol signaling pathway has also been studied for some FMPP mutations. The basal inositol phosphate levels of cells transfected with hLHR cDNA carrying Asp564Gly, Thr577Ile, Asp578Gly, Asp578Tyr, or Cys581Arg do not differ from those of cells transfected with the wild-type hLHR cDNA (Yano et al., 1994; Kosugi et al., 1995; Metzger and Chan, unpublished observations). On the other hand, cells expressing mutant hLHRs with the mutations Met398Thr, Met571Ile, or Ala572Val were found to have slightly increased basal levels of inositol phosphate (Kosugi et al., 1995; Yano et al., 1995, 1996). At concentrations of 10^{-9} mol/L or greater, hCG causes a concentration-dependent increase in inositol phosphate levels in cells transfected with either the wild-type hLHR or the mutant hLHRs with the exception of those carrying the Asp578Tyr or Cys581Arg mutation. In such cases the mutant hLHRs are unresponsive to hormone stimulation (Metzger and Chan, unpublished observations). The maximal hCG-stimulated inositol phosphate level achieved is significantly reduced in cells expressing mutant hLHRs carrying the Asp564Gly,

Asp578Gly, Ala572Val, or Met398Thr mutations (Yano et al., 1994, 1995, 1996; Metzger and Chan, unpublished observations) while the Thr577Ile and Met571Ile mutations do not affect the maximal hCG-stimulated inositol phosphate level achieved by the transfected cells (Kosugi et al., 1995). It is clear from the above discussion that the majority of the FMPP mutations do not have any significant effect on the phosphatidylinositol signaling activity of the hLHR. Therefore, it is fair to speculate that FMPP is not caused by an alteration in the phosphatidylinositol signaling pathway.

The role of the different mutations in causing constitutive activation of the hLHR is still unclear. A model of the probable arrangement of the seven TM helices in G-protein-coupled receptors places the most frequently mutated amino acid in FMPP, i.e., Asp578, in the middle of the TM VI (MaloneyHuss and Lybrand, 1992; Probst et al., 1992). The fact that the majority of the constitutively activating mutations of the hLHR as well as that of a number of G-protein-coupled receptors are produced by altering amino acids in the TM VI and the third cytoplasmic loop (Kjelsberg et al., 1992; Parma et al., 1995) suggests that these parts of the molecule may be important in relaying a signal from the ligand binding pocket to the intracellular face of the receptor. Asp578 is conserved in all the glycoprotein hormone receptors, but is not found in any of the other known G-protein-coupled receptors (Probst et al., 1992). However, amino acids occupying the equivalent position in some of these G-protein-coupled receptors are also postulated to be involved in signal propagation (Neitz et al., 1991; Trumpp-Kallmeyer et al., 1992). In the inactive state, amino acid residues of the hLHR involved in activating mutations, in particular Asp578, are thought to interact with neighboring amino acids through electrostatic or hydrogen bonds, thus providing a conformational constraint. When agonist binds to the hLHR, these bonds are disrupted resulting in a conformational change. Substitution of the polar Asp578 with Gly, which is incapable of forming electrostatic or hydrogen bonds, would permit the hLHR to assume a partially active conformation at all times independent of ligand. This substitution appears to have little effect on agonist affinity which is primarily determined by binding to the extracellular domain. If this speculation is correct, a corresponding change in the amino acid residue involved in bond formation with Asp578 or the other amino acids, which give rise to activating mutation, will also lead to conformational changes resulting in constitutive activation of the hLHR signaling pathway. Since α helices contain 3.6

amino acid residues per helical turn, the mutations so far identified in TM VI of hLHR in FMPP can be aligned precisely on the same side of the helix (Table 2). Although the significance of this finding is not obvious now, the structure-activity relationship of the different FMPP mutations could be resolved in the future through molecular modeling.

Genotype-Phenotype Correlation

While the mutant hLHRs are present on the Leydig cell from birth, the age at which signs of puberty develop in boys with FMPP is around 3 years. This delay in phenotypic expression may depend on the relative expression of other genes critical for Leydig cell maturation as well as the extent to which the mutant allele is expressed as protein. The delay might also depend on the relative activity of the activated receptor. As discussed in the previous paragraphs, there are differences in basal activities, responsiveness to hormone stimulation, agonist affinity, and receptor expression among the different mutant hLHRs. The relationship between the different genotypes and the clinical expression of the diseases is largely unknown, with one exception. Examination of the basal activity of the mutated hLHRs in the cAMP signaling pathway indicates close correlation between genotype and phenotype in at least one of the mutations. Expression of the mutated hLHR carrying the Asp578Tyr mutation in HEK 293 cells produces the highest basal levels of cAMP that are responsive to hCG-stimulation (Laue et al., 1995).

Table 2. Alignment of Amino Acid Residues with Respect to Helical Turns in TM VI of hLHR

Normal	Mutation	Position No.	Helix Turn No.
Met571	Ile	1	0
Ala572	Val	2	0
Ile575	Leu	5	1
Thr577	Ile	7	2
Asp578	Gly/Tyr	8	2
Cys581	Arg	11	3
Ile585		15	4
Ala589		19	5
Ala593		23	6

Note: Position of amino acids in TM VI of hLHR as described by Minegishi et al., 1990.

Both children identified with this mutation showed evidence of pubertal development by 12 months of age (Clark and Clarke, 1995; Laue et al., 1995), which is earlier than the other FMPP patients. Thus, it appears that this genotype correlates with a phenotype of earlier clinical expression of the disease.

FUTURE DIRECTION

So far, only two mutations of the $G_s\alpha$ have been detected in patients with MAS. Both mutations cause the substitution of Arg201 by another amino acid. Alteration of other amino acids of the $G_s\alpha$ in MAS has not been reported. The recent identification of Ala366Ser substitution in the $G_s\alpha$ of two patients with testotoxicosis and pseudohypoparathyroidism type Ia (Iiri et al., 1994) suggests that mutation of other amino acid residues of the $G_s\alpha$ might also result in constitutive activation of G_s and adenylyl cyclase. Therefore, investigation of the molecular genetics of $G_s\alpha$ in the other MAS patients in whom Arg201 and Ala366 are not mutated might lead to the definition of further $G_s\alpha$ amino acid substitutions resulting in the development of the disease. Similarly, there are a number of FMPP patients in whom no mutation in the LHR has been identified. In such cases, mutation in the non-coding sequence of the LHR gene or in other gene(s) in the LH response pathway might be responsible for the development of precocious puberty. Besides mutations of 21-hydroxylase, the $G_s\alpha$ subunit and the LHR, there are probably other genes the mutation of which gives rise to LHRH-independent precocious puberty. A recent report identifies the interaction of high concentrations of thyrotropin (TSH) with the FSH receptor as a pathogenic mechanism of precocious puberty in children with severe juvenile hypothyroidism (Anasti et al., 1995). Presumably mutation giving rise to either elevated secretion of TSH or constitutive activation of the FSH receptor would lead to precocious puberty. Only one activating mutation of the FSH receptor has so far been identified. This mutation leads to ligand-independent constitutive activation of the FSH receptor and was implicated to have caused autonomous sustainment of spermatogenesis in a hypophysectomized man (Gromoll et al., 1996). Thus molecular genetic studies of patients with LHRH-independent precocious puberty besides those with congenital adrenal hyperplasia (21-hydroxylase deficiency), MAS, and FMPP would help further define the genetic cause of this abnormality.

In vitro studies of the α_{1B}-adrenergic receptor led to the prediction of human disease due to activating mutations of G-protein-coupled receptor (Kjelsberg et al., 1992). Subsequently, such activating mutations have been discovered in retinitis pigmentosa (Robinson et al., 1992), FMPP (Shenker et al., 1993a), hyperfunctioning thyroid adenomas (Parma et al, 1993), familial hyperthyroidism (Duprez et al., 1994), Jansen-type metaphyseal chondrodysplasia (Schipani et al., 1995), autosomal dominant hypocalcemia (Pollack et al., 1995), and hypoparathyroidism (Baron et al., 1996). These discoveries suggest the hypothesis that all G-protein-coupled receptors are intrinsically vulnerable to activating mutations, which would have wide-reaching medical implications. For every G-protein-coupled receptor, it would be appropriate to ask, "What disorder would result if this receptor were constitutively activated by mutation?" And for every unexplained autosomal dominant disorder or endocrine neoplasm, it would be reasonable to consider, "Could this disorder result from the constitutive activation of a G-protein-coupled receptor?"

The activating mutations offer important clues to the question of how hormones activate receptors. Presumably, these mutations recreate the 3-dimensional surfaces that are normally induced by hormone binding and that lead in turn to G-protein binding and signal transduction. In the case of the LH receptor, all but one of the activating mutations fall within a 40-amino acid stretch from amino acid 542 to 581. This observation will focus future research (using computer modeling, site-directed mutagenesis, and chimeric receptors) on this critical region of the receptor.

The discovery of activating mutations of the G-protein-coupled receptors offers new opportunities to improve the treatment of these disorders. For example, "negative antagonists" (or "inverse agonists") have been proposed that induce the mutant receptor back into their resting, inactive conformation. For disorders that are suitable candidates for gene therapy, the discoveries focus research on how to prevent transcription or translation of the mutant allele. Thus, the insights into pathogenesis during the past several years are yielding new strategies to improve the treatment of these conditions in the future.

ACKNOWLEDGMENTS

W.Y.C. was supported in part by NIH grant HD31553.

REFERENCES

Anasti, J.N., Flack, M.R., Froehlich, J., Nelson, L.M., & Nisula, B.C. (1995). A potential novel mechanism for precocious puberty in juvenile hypothyroidism. J. Clin. Endocrinol. Metab. 80, 276-279.

Baron, J., Winer, K., Yanovski, J.A., Cunningham, A.W., Laue, L., Zimmerman, D., & Cutler, G.B., Jr. (1996). Mutations in the Ca^{2+}-sensing receptor gene cause autosomal dominant and sporadic hypoparathyroidism. Hum. Mol. Genet. 5, 601-606.

Boepple, P.A., Crowley, W.F., Jr., Albanese, C., & Jameson, J.L. (1994). Activating mutations of the LH receptor in sporadic male gonadotropin-independent precocious puberty. Endo. Soc. 76, A1176.

Brook, C.G.D. (1995). Precocious puberty. Clin. Endocrinol. 42, 647-650.

Clark, P.A., & Clarke, W.L. (1995). Testotoxicosis. An unusual presentation and novel gene mutation. Clin. Pediatr. 34, 271-274.

Cocco, S., Meloni, A., Marini, M.G., Cao, A., & Moi, P. (1996). A missense (T577I) mutation in the luteinizing hormone receptor gene associated with familial male-limited precocious puberty. Hum. Mut. 7, 164-166.

Cutler, G.B., Jr. (1992). Precocious puberty. In: Medicine for the Practicing Physician (Hurst, J.W., ed.), 3rd edn. pp. 577-581. Butterworth, Boston.

Cutler, G.B., Jr. (1993). Overview of premature sexual development. In: Sexual Precocity: Etiology, Diagnosis, and Management (Grave, G.D., & Cutler, G.B., Jr., eds.), pp. 1-10. Raven Press, New York.

Dumont, J.E., Jauniaux, J.C., & Roger, P.P. (1989). The cyclic AMP-mediated stimulation of cell proliferation. Trends Biochem. Sci. 14, 67-71.

Dunkel, L., Alftham, H., Stenman, U.-H., Selstam, G., Roseberg, S., & Albertsson-Wikland, K. (1992). Developmental changes in 24-hour profiles of luteinizing hormone and follicle-stimulating hormone from prepuberty to mid-stages of puberty in boys. J. Clin. Endocrinol. Metab. 74, 890-897.

Duprez, L., Parma, J., Van Sande, J., Allgeier, A., Leclere, J., Schvartz, C., Delisle, M.-J., Decoulx, M., Orgiazzi, J., Dumont, J., & Vassart, G. (1994). Germ-line mutations in the thyrotropin receptor gene cause non-autoimmune autosomal dominant hyperthyroidism. Nat. Genet. 7, 396-401.

Evans, B.A.J., Bowen, D.J., Smith, P.J., Clayton, P.E., & Gregory, J.W. (1996). A new point mutation in the luteinizing hormone receptor gene in familial and sporadic male-limited precocious puberty: Genotype does not always correlate with phenotype. J. Med. Genet. 33, 143-147.

Gromoll, J., Simoni, M., & Nieschlag, E. (1996). An activating mutation of the follicle-stimulating hormone receptor autonomously sustains spermatogenesis in a hypophysectomized man. J. Clin. Endocrinol. Metab. 81, 1367-1370.

Grumbach, M.M., & Styne, D.M. (1992). Puberty: Ontogeny, neuroendocrinology, physiology, and disorders. In: Williams Textbook of Endocrinology (Wilson, J.D., & Foster, D.W., eds.), 8th edn. pp. 1139-1208. W.D. Saunders, New York.

Happle, R. (1986). The McCune-Albright syndrome: A lethal gene surviving by mosaicism. Clin. Genet. 29, 321-324.

Holland, F.J. (1991). Gonadotropin-independent precocious puberty. Endocrinol. Metab. Clin. North Amer. 20, 191-210.

Iiri, T., Herzmark, P., Nakamoto, J.M., Van Dop, C., & Bourne, H.R. (1994). Rapid GDP release from $G_s\alpha$ in patients with gain and loss of endocrine function. Nature 371,164-168.

Kakar, S.S., Musgrove, L.C., Devor, D.C., Sellers, J.C., & Neill, J.D. (1992). Cloning, sequencing, and expression of human gonadotropin releasing hormone (GnRH) receptor. Biochem. Biophys. Res. Commun. 189, 289-295.

Kawate, N., Kletter, G.B., Wilson, B.E., Netzloff, M.L., & Menon, K.M.J. (1995). Identification of constitutively activating mutation of the luteinizing hormone receptor in a family with male limited gonadotrophin independent precocious puberty (testotoxicosis). J. Med. Genet. 32, 553-554.

Kelch, R.P., Hopwood, H.J., Saunder, S., & Marshall, J.C. (1985). Evidence for decreased secretion of gonadotropin-releasing hormone in pubertal boys during short-term testosterone treatment. Pediatr. Res. 19, 112-117.

Kjelsberg, M.A., Cotecchia, S., Ostrowski, J., Caron, M.G., & Lefkowitz, R.J. (1992). Constitutive activation of the α_{1B}-adrenergic receptor by all amino acid substitutions at a single site. J. Biol. Chem. 267, 1430-1433.

Kletter, G.B., & Kelch, R.P. (1993). Disorders of puberty in boys. Endocrinol. Metab. Clin. North Amer. 22, 455-477.

Kosugi, S., Van Dop, C., Geffner, M.E., Rabl, W., Carel, J.C., Chaussain, J.L., Mori, T., Merendino, J.J., & Shenker, A. (1995). Characterization of heterogeneous mutations causing constitutive activation of the luteinizing hormone receptor in familial male precocious puberty. Hum. Mol. Genet. 4, 183-188.

Kraaij, R., Post, M., Kremer, H., Milgrom, E., Epping, W., Brunner, H.G., Grootegoed, J.A., & Themmen, A.P.N. (1995). A missense mutation in the second transmembrane segment of the luteinizing hormone receptor causes familial male-limited precocious puberty. J. Clin. Endocrinol. Metab. 80, 3168-3172.

Kremer, H., Mariman, E., Otten, B.J., Moll, G.W., Jr., Stoelinga, G.B.A., Wit, J.M., Jansen, M., Drop, S.L., Faas, B., Ropers, H.H., & Brunner, H.G. (1993). Cosegregation of missense mutations of the luteinizing hormone receptor gene with familial male-limited precocious puberty. Hum. Mol. Genet. 2, 1779-1783.

Landis, C.A., Masters, S.B., Spada, A., Pace A.M., Bourne, H.R., & Vallar, L. (1989). GTPase inhibiting mutations activate the α chain of G_s and stimulate adenylyl cyclase in human pituitary tumours. Nature 340, 692-696.

Latronico, A.C., Anasti, J., Arnhold, I., Mendonca, B.B., Domenice, S., Albano, M.C., Zachman, K., Wajchenberg, B.L., Zachman, K., & Tsigos, C. (1995). A novel mutation of the luteinizing hormone receptor gene causing male gonadotropin-independent precocious puberty. J. Clin. Endocrinol. Metab. 80, 2490-2494.

Laue, L., Chan, W.Y., Hsueh, A.J.W., Kudo, M., Hsu, S.Y., Wu, S.M., Blomberg, L.A., & Cutler, G.B., Jr. (1995). Genetic heterogeneity of constitutively activating mutations of the human luteinizing hormone receptor in familial male precocious puberty. Proc. Natl. Acad. Sci. USA 92, 1906-1910.

Laue, L., & Rennert, O.M. (1995). Congenital adrenal hyperplasia: Molecular genetics and alternative approaches to treatment. Adv. Pediatr. 42, 113-143.

Laue, L., Wu, S.M., Kudo, M., Hsueh, A.J.W., Cutler, G.B., Jr., Jelly, D.H., Diamond, F.B., & Chan, W.Y. (1996). Heterogeneity of activating mutations of the human luteinizing hormone receptor in male-limited precocious puberty. Biochem. Mol. Med. 58, 192-198.

Levine, M. (1991). The McCune-Albright syndrome. N. Eng. J. Med. 325, 1738-1740.

Malchoff, C.D., Reardon, G., MacGillivray, D.C., Yamase, H., Rogol, A.D., & Malchoff, D.M. (1994). An unusual presentation of McCune-Albright syndrome confirmed by an activating mutation of the $G_s\alpha$-subunit from a bone lesion. J. Clin. Endocrinol. Metab. 78, 803-806.

MaloneyHuss, K., & Lybrand, T.P. (1992). Three-dimensional structure for the $\beta2$-adrenergic receptor protein based on computer modeling studies. J. Mol. Biol. 225, 859-871.

Minegishi, T., Nakamura, K., Takakura, Y., Miyamoto, K., Hasegawa, K., Ibuki, Y., & Igarashi, M. (1990). Cloning and sequencing of human LH/CG receptor cDNA. Biochem. Biophys. Res. Commun. 172, 1049-1054.

Müller, J., Kosugi, S., & Shenker, A. (1995). A severe, non-familial case of testoxicosis associated with a new mutation (Asp578 to Tyr) of the lutropin receptor (LHR) gene. Hormone Res. 44(Supp 1), 13 (Abstract).

Neitz, M., Neitz, J., & Jacobs, G.H. (1991). Spectral tuning of pigments underlying red-green color vision. Science 252, 971-974.

Oerter, K.E., Uriarte, M.M., Rose S.R., Barnes, K.M., & Cutler, G.B., Jr. (1990). Gonadotropin secretory dynamics during puberty in normal girls and boys. J. Clin. Endocrinol. Metab. 71, 1251-1258.

Parma, J., Duprez, L., Van Sande, J., Cochaux, P., Gervy, C., Mockel, J., Dumont, J., & Vassart, G. (1993). Somatic mutations in the thyrotropin receptor gene cause hyperfunctioning thyroid adenomas. Nature 365, 649-651.

Parma, J., Van Sande, J., Swillens, S., Tonacchera, M., Dumont, J., & Vassart, G. (1995). Somatic mutations causing constitutive activity of the thyrotropin receptor are the major cause of hyperfunctioning thyroid adenomas: Identification of additional mutations activating both the cyclic adenosine 3',5'-monophosphate and inositol phosphate-Ca^{2+} cascades. Mol. Endocrinol. 9, 725-733.

Pollack, M.R., Brown, E.M., Estep, H.L., McLaine, P.N., Kifor, O., Park, J., Hebert, S.C., Seidman, C.E., & Seidman, J.G. (1994). Autosomal dominant hypocalgaemia caused by a Ca^{2+}-sensing receptor gene mutation. Nature Genet. 8, 303-307.

Probost, W.C., Snyder, L.A., Schuster, D.l., Brosius, J., & Sealfon, S.C. (1992). Sequence alignment of the G-protein coupled receptor superfamily. DNA Cell. Biol. 11, 1-20.

Robinson, P.R., Cohen, G.B., Zhukovsky, E.A., & Oprian, D.D. (1992). Constitutively active mutants of rhodopsin. Neuron 9, 719-725.

Rosenthal, S.M., Grumbach, M.M., & Kaplan, S.L. (1983). Gonadotropin-independent familial sexual precocity with premature Leydig and germinal cell maturation (familial testotoxicosis): Effects of a potent luteinizing hormone-releasing factor agonist and medroxyprogesterone-acetate therapy in four cases. J. Clin. Endocrinol. Metab. 57, 571-579.

Schedewie, H.K., Reiter, E.O., Beitins, I.Z., Seyed, S., Wooten, V.D., Jimenez, J.F., Aiman, E., De Vane, G.W., Redman, J.F., & Elders, J.M. (1981). Testicular Leydig cell hyperplasia as a cause of familial sexual precocity. J. Clin. Endocrinol. Metab. 52, 271-278.

Schipani, E., Kruse, K., & Juppner, H. (1995). A constitutively active mutant PTH-PTHrP receptor in Jansen-type metaphyseal chondrodysplasia. Science 268, 98-100.

Schwindinger, W.F., Francomano, C.A., & Levine, M.A. (1992). Identification of a mutation in the gene encoding the subunit of the stimulatory G protein of adenylyl cyclase in McCune-Albright syndrome. Proc. Natl. Acad. Sci. USA 89, 5152-5156.

Schwindinger, W.F., & Levine, M.A. (1993). McCune Albright Syndrome. Trends Endocrinol. Metab. 4, 238-242.

Seeburg, P.H., & Adelman, J.P. (1984). Characterization of cDNA for precursor of human luteinizing hormone releasing hormone. Nature 311, 666-668.

Shenker, A., Chanson, P., Weinstein, L.S., Speigel, A.M., Lomri, A., & Marie, P.J. (1995). Osteoblastic cells derived from isolated lesions of fibrous dysplasia contain activating somatic mutations of the $G_s\alpha$ gene. Hum. Mol. Genet. 4, 1675-1676.

Shenker, A., Laue, L., Kosugi, S., Merendino, J.J., Jr., Minegishi, T., & Cutler, G.B., Jr. (1993a). A constitutively activating mutation of the luteinizing hormone receptor in familial male precocious puberty. Nature 365, 652-654.

Shenker, A., Weinstein, L.S., Moran, A., Pescovitz, O.H., Charest, N.J., Boney, C.M., van Wyk, J.J., Merino, M.J., Feuillan, P.P., & Spiegel, A.M. (1993b). Severe endocrine and nonendocrine manifestations of the McCune-Albright syndrome associated with activating mutations of stimulatory G protein G_s. J. Pediatr. 123, 509-518.

Shenker, A., Weinstein, L.S., Sweet, D.E., & Spiegel, A.M. (1994). An activating $G_s\alpha$ mutation is present in fibrous dysplasia of bone in the McCune-Albright syndrome. J. Clin. Endocrinol. Metab. 79, 750-755.

Sklar, C.A., Conte, F.A., Kaplan, S.L., & Grumbach, M.M. (1981). hCG-secreting pineal tumor. Relation to pathogenesis and sex limitation of sexual precocity. J. Clin. Endocrinol. Metab. 53, 656-660.

Trumpp-Kallmeyer, S., Hoflack, J., Bruinvels, A., & Hibert, M. (1992). Modeling of G protein-coupled receptors: Application to dopamine, adrenaline, serotonin, acetylcholine, and mammalian opsin receptors. J. Med. Chem. 35, 3448-3462.

Weinstein, L.S., Shenker, A., Gejman, P.V., Merino M.J., Friedman, E., & Spiegel A.M. (1991). Activating mutations of the stimulatory G protein in the McCune-Albright syndrome. N. Eng. J. Med. 325, 1688-1695.

White, P.C. (1989). Analysis of mutations causing steroid 21-hydroxylase deficiency. Endocr. Res. 15, 239-256.

White, P.C., Grossberger, D., Onufer, B.J., Chaplin, D.D., Maria, I.N., Dupont, B., & Strominger, J.L. (1985). Two genes encoding steroid 21-hydroxylase are located near the fourth component of complement. Proc. Natl. Acad. Sci. USA 82:1089-1096.

White, P.C., & New, M.I. (1988). Molecular genetics of congenital adrenal hyperplasia. Bailleres Clin. Endocrinol. Metab. 2, 941-965.

White, P.C., Vitek, A., Dupont B., & New, M.I. (1988). Characterization of frequent deletions causing steroid 21-hydroxylase deficiency. Proc. Natl. Acad. Sci. USA 85, 4436-4440.

Yano, K., Hidaka, A., Saji, M., Polymeropoulos, A., Okuno, A., Kohn, L.D., & Cutler, G.B., Jr. (1994). A sporadic case of male-limited precocious puberty has the same constitutively activating point mutation in luteinizing hormone/choriogonadotropin receptor as familial cases. J. Clin. Endocrinol. Metab. 79, 1818-1823.

Yano, K., Kohn, L.D., Saji, M., Kataoka, N., Okuno, A., & Cutler, G.B., Jr. (1996). A case of male-limited precocious puberty caused by a point mutation in the second transmembrane domain of the luteinizing hormone choriogonadotropin receptor gene. Biochem. Biophys. Res. Commun. 220, 1036-1042.

Yano, K,, Saji, M., Hidaka, A., Moriya, N., Okuno, A., Kohn, L.D., Cutler, G.B., Jr. (1995). A new constitutively activating point mutation in the luteinizing hormone/chorionic gonadotropin receptor gene in cases of male-limited precocious puberty. J. Clin. Endocrinol. Metab. 80, 1162-1168.

Zung, A., Chalew, S.A., Schwindinger, W.F., Levine, M.A., Phillip, M., Jara, A., Counts, D.R., & Kowarski, A.A. (1995). Uriniary cyclic adenosine 3',5'-monophosphate response in McCune-Albright syndrome: Clinical evidence for altered renal adenylate cyclase activity. J. Clin. Endocrinol. Metab. 80, 3576-3581.

Chapter 7

Two Genes–One Disease: The Molecular Basis of Congenital Nephrogenic Diabetes Insipidus

WALTER ROSENTHAL, ALEXANDER OKSCHE,

and DANIEL G. BICHET

Advances in Molecular and Cellular Endocrinology
Volume 2, pages 143-167.
Copyright © 1998 by JAI Press Inc.
All rights of reproduction in any form reserved.
ISBN: 0-7623-0292-5

INTRODUCTION

Congenital nephrogenic diabetes insipidus (NDI) is a rare disorder of the kidney, characterized by the failure to concentrate urine despite normal or elevated levels of the antidiuretic hormone arginine-vasopressin. After molecular cloning of the V2 receptor and mapping of the gene to the X-chromosome, about 70 different mutations in the V2 receptor gene have been described in families with X-linked NDI. The majority of mutations are missense and nonsense mutations; frameshift mutations due to insertion or deletion of bases have also been observed. The mutations are almost evenly distributed within the receptor molecule, and only few recurrent mutations have evolved. Functional characterization of V2 receptor mutants revealed the following defects: a) defect of processing, b) defect of transport to the cell surface, c) defect of ligand binding, and d) defect of G protein coupling/activation. In a minority of NDI families, an autosomal-recessive mode of inheritance has been found. This form of congenital NDI is caused by mutations of the recently identified aquaporin-2 gene, located on chromosome 12. So far, six different mutations of the aquaporin-2 gene have been reported: five missense mutations and a single base deletion causing a frameshift. The functional impairment of mutants appears to be mainly a consequence of impaired cellular routing. The identification of two genes responsible for congenital nephrogenic diabetes insipidus nephrogenic diabetes insipidus opens the door for the development of therapeutic strategies based on gene transfer.

VASOPRESSIN RECEPTORS

Nonapeptides of the vasopressin family are the key regulators of water homeostasis in amphibia, reptiles, birds, and mammals. Since these peptides reduce urinary output, they are also referred to as antidiuretic hormones. The antidiuretic hormone in man and most mammals is 8-arginine vasopressin. A vasopressin precursor is synthesized in the hypothalamic supraoptic and paraventricular nuclei. Proteolytic processing takes place in the producing neurons and during axonal transport to the posterior pituitary, where the hormone is stored in vesicles. Exocytotic release is stimulated by minute increases in serum osmolality (hypernatremia, osmotic regulation) and by more pronounced decreases of extracellular fluid (hypovolemia, nonosmotic regulation).

In the kidney, the molecular targets of vasopressin are V2 receptors expressed on the basolateral surface of principal cells in the collecting duct epithelium (Figure 1). The deduced primary structure of the human

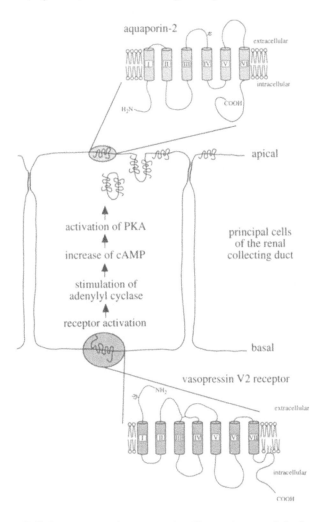

Figure 1. Cellular actions of vasopressin. Shown is a model of a principal cell. The transmembrane regions of the V2 receptor and aquaporin-2 (AQP-2) are numbered with roman figures (insets). The second extracellular loop of AQP-2 and the extracellular N-terminus of the V2 receptor are shown to be glycosylated; the intracellular C-terminus of the V2 receptors is shown to be modified by two fatty acid residues. Note the reverse representation of the V2 receptor in the plasma membrane and in the inset.

(Birnbaumer et al., 1992; Seibold et al., 1992) and rat (Lolait et al., 1992) V2 receptor confirmed ample biochemical evidence identifying them as members of the superfamily of G protein-coupled receptors (Figure 2). The receptors are positively coupled to adenylyl cyclases by the stimulatory G protein G_s. Although other signal pathways have not been rigor-

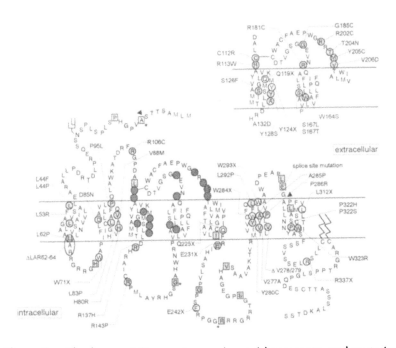

Figure 2. The human V2 receptor: amino acid sequence and mutations associated with X-linked NDI. Arrowheads indicate the positions of introns in the genomic DNA. Posttranslational modifications are depicted as in Figure 1. Mutations: missense mutations are depicted by hatched circles, nonsense mutations by open circles, and deletions and insertions causing a frameshift by hatched squares at the predicted location of the mutation. An asterisk at the hatched squares indicates two frameshift mutations within a codon. In-frame deletions are shown at the junction of the first transmembrane domain and the first intracellular loop (ΔLAR62-64) and in the sixth transmembrane domain (ΔV278/279, deletion of V 278 or 279). A splice site mutation is depicted in the third extracellular loop. The mutations in the third and fourth transmembrane domains and the second extracellular loop (indicated by dotted circles) are described in the inset. For a description of mutations, see Fujiwara et al., 1995 and references cited therein; references in Table 2; Tajima et al., 1996.

ously excluded, the hormone-induced rise in intracellular cAMP and the subsequent activation of cAMP-dependent protein kinase (PKA) appear to be the main mediators of the cellular response (for review, see Breyer and Ando, 1994). The vasopressin-induced increase in water permeability permits reabsorption of water from the lumen of the collecting duct, thereby increasing urine osmolality and reducing urinary volume. Whereas the main site of vasopressin action in mammals is clearly the renal collecting duct, vasopressin also activates the $Na^+:K^+:2Cl^-$-cotransporter in the ascending loop of Henle and thereby maintains the driving force for reabsorption of water, i.e., the osmotic gradient between the tubular fluid and the medullary interstitium. This effect, which seems to be species-dependent, is also mediated by V2 receptors.

V2 receptors are apparently also involved in extrarenal responses to vasopressin. This assumption is based on pharmacological and genetic grounds. Pharmacological evidence is mainly derived from the effects of the synthetic vasopressin analogue, desmopressin, a highly specific V2 receptor agonist which lacks pressor effects. Desmopressin not only causes antidiuresis but also a decrease in the peripheral resistance (vasodilatory response) and an increase in the plasma concentrations of coagulation factor VIIIc and of von Willebrand factor which promotes platelet adhesion and serves as a carrier for factor VIIIc; it also enhances plasma levels of tissue plasminogen activator (tPA). The increase in factor VIIIc and von Willebrand factor (coagulation response) is the rationale for the widespread clinical use of desmopressin for the treatment of bleeding disorders such as mild forms of hemophilia and von Willebrand disease. The vasodilatory response may be observed as a side effect (flush) of desmopressin intranasally administered to patients with central diabetes insipidus (replacement therapy). The genetic evidence for extrarenal V2 receptors comes from the observation that patients with X-linked NDI, caused by a defect in the V2 receptor gene (see below), lack the vasodilatory and the coagulation response (Bichet et al., 1988). The structural search for extrarenal V2 receptors has been elusive.

Other, mostly extrarenal effects of vasopressin require receptors distinct from the V2 receptor. Vasopressin V1a receptors are widely distributed. Expressed on vascular smooth muscle cells and on hepatocytes, they mediate vasopressin-induced vasoconstriction and glycogenolysis, respectively. Stimulation of vasopressin V1b receptors, present on corticotrophs in the anterior pituitary, promotes the release of corticotropin. Both V1a and V1b receptors activate the Gq/phospholipase C system, possess the structural key features of G protein-coupled receptors (see Figure 2) and show considerable amino acid identities with the V2 receptor (Sugimoto et al., 1994; Thibonnier et al., 1994). In contrast, a recently identified dual angiotensin II/vasopressin receptor, expressed in the kidney and in other organs, shows an atypical membrane topology (extracellular C-terminus) and lacks sequence homology with the other vasopressin receptors (Ruiz-Opazo et al., 1995). The functional significance of this receptor, which couples to the G_s/adenylyl cyclase system, is not known.

AQUAPORINS

Whereas V2 receptors represent the first cellular component of the signal transduction cascade activated by vasopressin, the ultimate cellular targets of the hormone in the collecting duct are aquaporins (AQPs) (for recent reviews, see Knepper, 1994; Agre et al., 1995). By forming water-selective channels, AQPs considerably increase the water permeablilty of plasma membranes and are crucial for rapid water transport across tight epithelia. AQPs belong to the major intrinsic protein (MIP) family, named after a protein of the lens. Members of this protein family have been identified in plants, bacteria, yeast, insects, and vertebrates. Hydropathy analysis of their deduced amino acid sequences and structural studies on the protein level indicate the presence of cytoplasmic amino and carboxy termini and six transmembrane α-helical domains (see Figures 1 and 3). Like the other members of the MIP family, aquaporins contain a NPA motif in both the first intracellular and the third extracellular loop. According to the hourglass model, which takes into account the inverse symmetry of the amino- vs. carboxy-terminal half of the aquaporins, these two loops dip into the membrane from opposite sites with the overlapping portions (NPA motifs) forming a narrow aqueous channel (Jung et al., 1994). Recent data suggest that in case of AQP-2 (see below), the second extracellular and second intracellular loops are in close proximity to the aqueous pore (Bai et al., 1996). Hydrodynamic studies and high-resolution electron microscopy have shown that AQPs are organized as homotetramers (reviewed in Agre et al., 1995).

AQP-1 (originally referred to as CHIP28) is a highly abundant protein in erythrocyte membranes (Preston and Agre, 1991). Within the nephron, AQP-1 is found in both apical and basolateral membranes of epithelial cells of the proximal tubule and of the descending loop of Henle (Sabolic

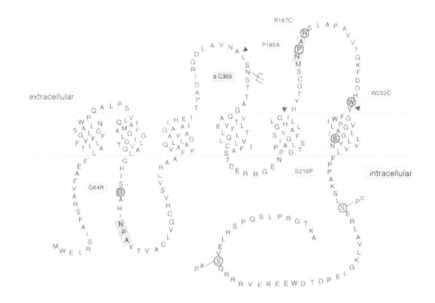

Figure 3. The human aquaporin 2: amino acid sequence and mutations associated with autosomal-recessive NDI. Arrows indicate the positions of introns in the genomic DNA. The NPA motifs in the first intracellular and third extracellular loops are boxed. Consensus sites for phoysphorylation by PKA (p^a) and protein kinase C (p^c) are depicted by open circles. Posttranslational modifications are depicted as in Figure 1. Mutations: missense mutations are depicted by hatched circles, and a single-base deletion by a line drawn between the last amino acid of the wild-type sequence and the first amino acid of the altered sequence. For a description of mutations, see Deen et al., 1994; Lieburg et al., 1995; Oksche et al., 1996a.

et al., 1992; Nielsen et al., 1993b). At these sites and probably in other epithelia AQP-1 may contribute to a constitutively rapid water transport. It is, however, unlikely to be crucial for this function, since patients lacking functional AQP-1 are phenotypically normal (Preston et al., 1994). In contrast to AQP-1, AQP-2, originally termed AQP-CD, has only been found in principal cells of the collecting duct (Fushimi et al., 1993). This observation, together with the subcellular distribution of AQP-2 (apical membrane and intracellular vesicles; see Figure 1; Nielsen et al., 1993a), and the finding that inactivating mutations are associated with a rare (autosomal-recessive) form of NDI (Deen et al., 1994) provide compelling evidence that AQP-2 is the vasopressin-regulated AQP. A common feature of AQP-1 and AQP-2 is their inhibition by mercury. In the case of AQP-2, cysteine 181, located in the third extracellular loop, was identified as the mercury-sensitive site (Bai et al., 1996). Another mercury-sensitive aquaporin, AQP-3, is mainly found in the basolateral membrane of principal cells (Ecelbarger et al., 1995). It shares this location with a mercury-insensitive aquaporin, AQP-4, which is also found in other organs (brain, stomach; Frigeri et al., 1995; Terris et al., 1995). AQP-3 and AQP-4 may play a role in the constitutively high water permeability of the basolateral membrane of principal cells. Another aquaporin, AQP-5, is not detectable in the kidney (Raina et al., 1995).

Whereas two aquaporins, AQP-2 and AQP-3, are regulated long-term by water restriction at the transcriptional level, only AQP-2 shows a fast hormonal regulation which does not require protein synthesis. According to the "shuttle hypothesis" originally proposed by Wade (Wade et al., 1981), vasopressin increases water permeability by exocytotic insertion of AQPs into the apical membrane (for recent reviews, see Handler, 1988; Breyer and Ando, 1994; Hays et al., 1994). This hypothesis is strongly supported by electron microscopy studies, which show a vasopressin-induced redistribution of AQP-2 from intracellular vesicles to the plasma membrane (Nielsen et al., 1995a; Sabolic et al., 1995). The time course of the translocation (observed within minutes) correlates well with that of the increase in water permeability in response to vasopressin.

At present, the mechanism of AQP-2 translocation is not understood. A first step towards elucidation of the events following the rise in cAMP and activation of PKA is the finding that vesicular proteins known to be involved in exocytosis are associated with AQP-2 containing vesicles. These include synaptobrevin II (or VAMP2) and Rab3 proteins (Jo et al., 1995; Liebenhoff et al., 1995; Nielsen et al., 1995b). In particular the identification of Rab3 proteins which are specifically linked to regulated exoctyosis suggests that the mechanism may be similar to the exocytotic process in neurons and other secretory cells.

Two aquaporins, AQP-2 and AQP-5, possess a PKA consensus phosphorylation site. After injection of *Xenopus* oocytes with rat AQP-2 cRNA, cAMP or forskolin, an activator of adenylyl cyclase, stimulate water permeability by about 50% (Kuwahara et al., 1995). This increase in water permeability is small compared to the one observed in the kidney (five-fold). Phosphorylation of AQP-2 on vesicles isolated from collecting ducts, however, does not influence water permeability of these vesicles (Lande et al., 1996). These data suggest that the intrinsic activity of AQP-2 is not substantially stimulated by PKA-catalyzed phosphorylation. It is therefore likely that the major role of PKA is to trigger translocation, either by phosphorylation of an as yet unidentified protein or by phosphorylation of AQP-2. The cAMP-dependency of the translocation process has recently been demonstrated in kidney epithelial cell lines overexpressing AQP-2 (Katsura et al., 1995; Valenti et al., 1995).

NEPHROGENIC DIABETES INSIPIDUS

The key symptoms of all forms of diabetes insipidus are polyuria and polydipsia. In general, the urine osmolality is lower in the urine than in the blood (< 300 mOsmol/kg). The daily excreted volume of urine ranges between 2.5 L/day in mild cases to 25 L/day in severe cases.

Central diabetes insipidus is caused by insufficient or no release of vasopressin from the posterior pituitary and is therefore treatable with desmopressin. Frequent causes are brain tumors and head injuries. Inherited central diabetes insipidus is rare. It derives from mutations in the vasopressin gene which disturb processing of the precursor (Rittig et al., 1996; for review see Schmale et al., 1993). NDI is caused by resistance of the kidney towards vasopressin. The diagnosis of NDI is mainly based on the failure of desmopressin to raise urine osmolality. Acquired and inherited forms of NDI have been described. Acquired causes include chronic renal diseases, potassium deficiencies, chronic hypercalcemia, systemic disorders (e.g., multiple myeloma, amyloidosis, Sjögren syndrome, and sarcoidosis), and drugs. Among the latter is Li$^+$, used for the treatment of bipolar manic-depressive illness, which frequently causes NDI. Other drugs known to cause NDI are the tetracylin antibiotic demeclocyclin and the volatile fluorocarbon anesthetic methoxyflurane. Congenital NDI typically shows an X-linked mode of inheritance (90% of cases); in a minority of patients (10%) an autosomal-recessive trait is observed. In contrast to patients with congenital central diabetes insipidus who present with symptoms during childhood or adolescence, patients with congenital NDI present with symptoms (fever, vomiting) shortly after birth. Because of the very early onset of congenital NDI, severe episodes of dehydration may cause mental retardation, dwarfism, and even death.

Mutations in the Vasopressin V2 Receptor Gene Are Responsible for X-Linked NDI

Isolation of the human gene encoding the V2 receptor and its chromo-somal assignment were two important steps towards identification of the molecular defects in patients with X-linked NDI. A 2 kb fragment con-tains the entire coding sequence on three exons which are separated by two small introns, 100 bp of 5' untranslated sequence with a putative TATA box and 460 bp of 3' untranslated sequence terminating at the poly-adenylation signal (Figure 4; Pan et al., 1992; Seibold et al., 1992). In agreement with data based on the expression of vasopressin-binding sites on hybrid cells harbouring defined regions of the X-chromosome (Jans et al., 1990), the gene was mapped to the telomeric region of the long arm of the X-chromsome (Xq28) by several groups (Lolait et al., 1992; van den Ouweland, 1992; Seibold et al., 1992). The region was indistinguishable from the NDI locus inferred from restriction fragment length polymor-phism (RFLP) analysis (Knoers et al., 1988).

The availability of the genomic sequence allowed the synthesis of primers suitable for amplification of the the entire transcriptional unit (2 kb; see Figure 4) or portions of it by polymerase chain reaction; genomic DNA obtained from blood samples of patients with X-linked NDI and their relatives was used as a template. The first mutation identified was a

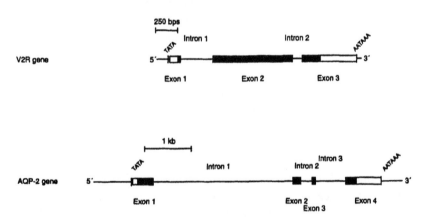

Figure 4.   Structures of the human V2 receptor (V2R) gene and the human aquaporin-2 (AQP-2) gene. Translated and untranslated portions of exons are shown as black or open bars, respectively. Introns are depicted as lines. TATA boxes are shown as vertical lines.

single base deletion causing a shift in the open reading frame found in a North American family with X-linked NDI (Rosenthal et al., 1992). Since then, approximately 70 different mutations have been discovered (see Figure 2 and references. in the legend to Figure 2), which are scattered throughout the protein. In addition, there is evidence for hot spots: some mutations have been found in two or more unrelated families. For example, mutations at serine 167 have been described in at least eight unrelated families (Bichet et al., 1994; Knoers et al., 1994; Oksche et al., 1994; Wildin et al., 1994). These data are reminiscent of those obtained from patients with late-onset autosomal-dominant retinitis pigmentosa. In one quarter of patients, the disease is caused by mutations in the light receptor rhodopsin. Here, too, many different mutations (100), spread throughout the coding region of the rhodopsin gene, have been found (Souied et al., 1994; Vaithinathan et al., 1994 and references cited therein).

The majority of the V2 receptor mutations are missense mutations, causing the exchange of a single amino acid; at present, 35 have been described in the literature (see Figure 2). In addition, 10 *Null* mutations (nonsense mutatons introducing a premature stop codon) and two inframe deletions have been reported. Moreover, 18 frameshift mutations are known. Finally, large deletions or gene rearrangements have been described in three families (not shown in Figure 2; see Bichet et al., 1994). Whereas deletions and insertions can be attributed to slipped mispairing during DNA replication favored by direct repeats, complementary repeats and symmetric sequences in the vicinity of the mutation, more than 50% of the single base substitutions can be explained by mutations in CpG dinucleotides, which are hot spots for genetic disease and which are relatively common in the human V2 receptor gene (Bichet et al., 1994).

Generally, X-linked NDI is a rare disease with an estimated prevalence of about four per million males and a carrier frequency of 7.4×10^{-6}; these figures are based on the number of patients with X-linked NDI known in Quebec (Bichet et al., 1992). In defined regions of North America, however, the prevalence is much higher. It is assumed that the patients in these regions are progeny of common ancestors. An example is the Mormon pedigree with its members residing in Utah (Utah families); this pedigree was originally described by Cannon (1955; see also Bichet et al., 1992). The "Utah mutation" is a missense mutation (L312X) predictive for a receptor which lacks transmembrane domain 7 and the intracellular C-terminus (Bichet et al., 1993). The largest known kindred with

X-linked NDI is the Hopewell family, named after the Irish ship Hopewell which arrived in Halifax, Nova Scotia, in 1761 (Bode and Crawford, 1969). Aboard the ship were members of the Ulster Scot clan, descendants of Scottish presbyterians who migrated to the Ulster Province of Ireland in the 17th century and left Ireland for the new world in the 18th century. Whereas families arriving with the first emigration wave settled in Northern Massachusetts in 1718, the members of a second emigration wave, passengers of the Hopewell, settled in Colchester County, Nova Scotia. According to the "Hopewell hypothesis" (Bode and Crawford, 1969), most NDI patients in North America are progeny of female carriers of the second emigration wave. This assumption is mainly based on the high prevalence of NDI among descendants of the Ulster Scots residing in Nova Scotia. In two villages with 2,500 inhabitants, 30 patients have been diagnosed, and the carrier frequency been estimated at 6% (Bichet et al., 1992). In consideration of the numerous mutations found in North American X-linked NDI families, the Hopewell hypothesis cannot be upheld in its originally proposed form. Yet the mutation found in members of the Hopewell pedigree (Hopewell mutation) is likely to be the most common mutation, responsible for X-linked NDI in more than 100 patients. It is a null mutation (W71X; Bichet et al., 1993; Holtzman et al., 1993) predictive for an extremely truncated receptor consisting of the extracelluar N-terminus, the first transmembrane domain and the N-terminal half of the first intracellular loop. Since the original carrier cannot be identified (Bichet et al., 1992), it is not clear whether the Hopewell mutation was brought to North America by Hopewell passengers or by other Ulster Scot immigrants.

The properties of missense and in-frame deletion mutant V2 receptors have been characterized in cells transiently or stably transfected with plasmids encoding the mutant receptors (Table 1). So far, major disturbances on the transcriptional level have not been observed. However, the following defects have been described: a) defect of processing, b) defect of transport to the cell surface, c) defect of ligand binding, and d) defect of G protein coupling/activation. A defect in protein processing has been found in the mutants L44P, W164S and in the frequently encountered mutants with substitutions at codon 167 (S167L, S167T) (Oksche et al., 1996b). The mutant proteins lack complex glycosylation and are expressed at low levels. This is most likely due to misfolding, resulting in retention in the endoplasmic reticulum and subsequent degradation. It is of note that the majority of mutations compromise the transport of the V2

Table 1. Functional Properties of V2 Receptor Mutants Associated With X-Linked NDI

Type of Mutation	Base Exchange	Amino Acid change	Ex-Location in V2R	mRNA Levels	Complex Glycosylation	Expression at Cell Surface	Ligand Affinity	Ability to Stimulate Adenylyl Cyclase	Reference
Missense	C498T	R113W	EL I	n.d.	n.d.	⇓	⇓	⇓	Birnbaumer et al., 1994
Missense	C612T	T181C	EL II	n.d.	yes	⇔	⇓	⇔	Pan et al, 1994
Missense	C675T	R202C	EL II	⇔	yes	COS cells: ⇓ CHO cells: ⇓	COS cells: ⇓ CHO cells: ⇓⇓⇓⇓	COS cells: n.d. CHO cells: ⇓⇓⇓⇓	Tsukaguchi et al., 1995a,c
Missense	A454C	Y128S	TMD II	n.d.	yes	⇔	⇓⇓⇓⇓	⇓⇓⇓⇓	Pan et al., 1994
Missense	C928G	**R286R**	TMD VI	n.d.	yes	⇔	⇓⇓⇓⇓	⇓⇓⇓⇓	Pan et al., 1994
Missense	C201T	L44F	TMD I	⇔	yes	n.d.	⇓⇓⇓⇓	⇓⇓⇓⇓	Oksche et al., 1996b
Missense	G481A	**R137H**	IL II	n.d.	n.d.	⇓	⇔	⇓⇓⇓⇓	Rosenthal et al., 1993
Missense	G499C	T143P	IL III	⇔	yes	⇓⇓⇓⇓	⇔	⇓	Tsukaguchi et al., 1995b,c
In-frame deletion	del1906-908	ΔV278/279	TMD VI	⇔	yes	⇓⇓⇓⇓	⇓⇓⇓⇓	⇓⇓⇓⇓	Tsukaguchi et al., 1995b,c
Missense	T202C	L44P	TMD I	⇔	no	n.d.	⇓⇓⇓⇓	n.d.	Oksche et al., 1996b
Missense	G562C	**W164S**	TMD IV	⇔	no	n.d.	⇓⇓⇓⇓	n.d.	Oksche et al., 1996b
Missense	C571T	**S167L**	TMD IV	⇔	no	n.d.	⇓⇓⇓⇓	n.d.	Oksche et al., 1996b
Missense	T570A	**S167T**	TMD IV	⇔	no	n.d.	⇓⇓⇓⇓	n.d.	Oksche et al., 1996b

Notes: Numbering of nucleotides is according to cDNA sequence (Birnbaumer et al., 1992).
Single letter code of amino acids is used; numbering of amino acids is according to the position in the protein.
IL, intracellular loop; EL, eaxtracellular loop; TMD, transmembrane domain; bold, substitutions of highly conserved amino acids among GPCRs;
n.d., not determined; ⇔, normal; ⇓, reduced; ⇓⇓⇓⇓, abolished.

receptor to the cell surface. This is a major obstacle to evaluating the ability of mutant receptors to bind the hormone and to transduce a signal. In the case of the in-frame deletion ΔV278/279 (lacking a valine at position 278 or 279 in the sixth transmembrane domain), impaired cell surface delivery may be the primary defect. This applies also to the missense mutation R143P, which additionally shows a slightly reduced ability to stimulate adenylyl cyclase. Missense mutations occuring in the first (R113W) and the second extracellular loop (R181C and R202C) have a lowered or non-detectable binding affinity for vasopressin, supporting the assumption that both loops contribute to the formation of the binding site for vasopressin (Kojro et al., 1993; Chini et al., 1995). In addition, the missense mutations, Y128S, in the second, and P286R, in the sixth transmembrane domain, cause a complete loss of vasopressin-binding activity, although the mutant receptors are expressed at the cell surface. A decrease or loss of binding activity may also be the primary defect of the mutant L44F (mutation at the junction of the extracellular N-terminus and the first transmembrane domain). This would explain why this mutant, although normally processed and therefore most likely expressed on the cell surface, is not capable of stimulating adenylyl cylase in response to vasopressin. The missense mutation R137H, in the second intracellular loop close to the third transmembrane domain, does not affect binding affinity for vasopressin but abolishes stimulation of adenylyl cyclase, indicating that residues of the second intracellular loop are indispensible for G protein coupling.

Null mutations (see Figure 2) are predictive for a truncated V2 receptor. They not only lack functional domains but are also expected to be retained and degraded within the cell. Surprisingly, the Q119X mutant (mutation in the second transmembrane domain) was shown to be expressed on the surface of COS cells. Recent studies with fusion proteins show that the first 71 amino acids (extracellular N-terminus, first transmembrane domain, and N-terminal portion of the first intracellular loop) contain the signals for both membrane insertion and correct orientation in the membrane (Schülein et al., 1996). Thus even the Hopewell mutant (W71X), which is the most extreme example of a naturally occurring truncated V2 receptor, should be properly inserted into the membrane of the endoplasmic reticulum; the fate of this mutant within the cell (processing, transport) has not been investigated.

In addition to the large number of mutations causing X-linked NDI, three mutations in the V2 receptor have been detected which do not appear to be causative for X-linked NDI. The A61V mutant (mutation in the first transmembrane domain) exhibits normal functional properties (cell surface expression, vasopressin binding, stimulation of adenylyl cyclase). It is therefore likely that X-linked NDI of the patient with this mutation is caused either by another mutation in the V2 receptor gene or by a mutation in a different gene. Like the A61V mutant, the mutant with the in-frame deletion del810–821 (deletion of four amino acids in the third cytoplasmic loop) shows normal functional properties. Originally this mutation was found in an allele also containing the R181C mutation (Pan et al., 1992). Subsequently, a family with an NDI allele harboring only the R181C mutation was found (Bichet et al., 1994). Thus the in-frame deletion del810–821 is probably a sequence variant which does not cause disease. Other, not yet functionally characterized missense mutations are also unlikely to cause X-linked NDI, as they either do not segregate with disease within an affected family (R64W) or have occurred on the NDI allele with a nonsense mutation which is likely to be causative for the disease (T7S, A147V; Bichet et al., 1994). Finally, several silent substitutions have been identified (Rosenthal et al., 1992; Bichet et al., 1994).

Usually patients with X-linked NDI do not show a progressive deterioration of kidney function. Principal cells may be able to degrade "toxic" mutant receptors fast enough to prevent accumulation. This is fundamentally different from late-onset autosomal-dominant retinitis pigmentosa. Here, accumulation of mutant rhodopsin within cells (e.g., Sung et al., 1993, 1994) causes a slowly progressing degeneration of the retina and thereby loss of vision in middle-aged individuals, probably by interference with the maturation of the wild-type rhodopsin by the mutant protein (Colley et al., 1995). This mechanism also explains the dominant inheritance of the disease.

Mutations in the AQP-2 Gene Are Responsible for Autosomal-Recessive NDI

Cloning of a rat cDNA encoding AQP-2 was accomplished in 1993 (Fushimi et al., 1993). The properties of AQP-2 suggested that it was the vasopressin-regulated water channel (see above). Subsequently, both the human cDNA (Sasaki et al., 1994) and gene (Uchida et al., 1994) were isolated. The latter comprises four exons distributed over 5 kb (see Figure 4). Their sizes are: exon 1, 454 bp; exon 2, 165 bp; exon 3, 81 bp; exon 4, 761 bp. Exon 1 contains 94 bp of untranslated sequence and the start codon, exon 4 551 bp of untranslated sequence and the polyadenylation signal. The sizes of introns 1, 2, and 3 are ~ 2.9, ~0.25 and ~0.7 kb, respectively. A TATA box is found 123 bp upstream of the start codon. Major transcription initiation sites are 30 and 31 bases downstream of the TATA box. Interestingly, the 5' flanking region contains—among other regulatory motifs—a consensus sequence for a cAMP-responsive element 314 bp upstream of the start codon (not shown). This may explain the increase in the transcriptional rate in response to water restriction (increase in vasopressin, see above).

The human AQP-2 gene was mapped to the long arm of chromsome 12 (12q13; Deen et al. 1994; Sasaki et al., 1994); this location is consistent with a role for the gene in autosomal-recessive NDI. Proof for such a role was provided by the demonstration of inactivating mutations in the gene in patients with autosomal-recessive NDI (Deen et al., 1994). To date six different mutations of the AQP-2 gene associated with autosomal-recessive NDI in five families have been reported—five missense mutations and a single base deletion causing a frameshift (see Figure 3; Deen et al., 1994; van Lieburg et al., 1994; Bichet et al., 1995; Oksche et al., 1996a). The first patient analyzed turned out to be a compound heterozygote, having inherited the R187C mutation from the father and the S216P mutation from the mother (Deen et al., 1994). In five subsequently analyzed families, the affected individuals were found to be homozygous for the mutations G64R, 369delC, P 185A, R187C, or W202C. Since the parents of these patients are consanguineous, they are likely to be heterozygous for the same mutation.

The mutants G64R, R187C, and S216P, when expressed in *Xenopus* oocytes, are not functional (Deen et al., 1994; van Lieburg et al., 1994). Interestingly, the coexpression of these mutants and wild type AQP-2 in oocytes does not affect the function of the latter. Although the system is rather artificial and allows only short-term expression of proteins, this finding may be regarded as evidence for the assumption that the mutant AQP-2 is nontoxic, in contrast to mutant rhodopsins causing autosomal-dominant retinitis pigmentosa (see above). The data may explain the recessive nature of NDI caused by AQP-2 defects. Further analysis of the the mutants G64R, R187C, and S216P in the oocyte system showed that they are partially expressed as immature proteins and retarded in the endoplasmic reticulum (Deen et al., 1995). Surface expression was greatly diminished or hardly detectable. Thus the functional impairment of these mutants appears to be a consequence of impaired cellular routing.

Although the other mutants have not been characterized, they are unlikely to be functional. The deletion of the nucleotide at position 369 (369 delC) results in the generation of a sequence of eight missense amino acids followed by premature termination. Thus the mutant lacks the C-terminal half of the protein. The mutant P185C changes the NPA motif in the third extracellular loop to NAA (see Figure 3). As pointed out above, the NPA motif is found in all members of the MIP family and is believed to form the aqueous pore together with the NPA motif in the first intracellular loop. Underlying the mutant W202C is a G to T transversion (G606T) within a 5' donor splice consensus site, adjacent to the invariable GT motif (see Figure 3). Although the invariant GT motif of the 5' splice donor site is unaltered, it is possible that the mutation causes aberrant splicing due to exon skipping or activation of cryptic splice sites, similar to G to T transversions at the corresponding position in other genes (see references in Oksche et al., 1996a). It is also feasible that the exchange of tryptophan by cysteine at position 202 profoundly affects the function of AQP-2, since the tryptophan is conserved in all human aquaporins.

PERSPECTIVES

Over the last few years it has become clear that congenital NDI is caused by an inactivating mutation of a G protein-coupled receptor (V2 receptor) or a water channel (AQP-2). While mutations in the V2-receptor gene are responsible for X-linked NDI, mutations in the AQP-2 gene account for autosomal-recessive NDI. The time of onset of the disease (shortly after birth) and the clinical symptoms do not differ between the two forms. However, the two forms can be distinguished by clinical testing: whereas desmopressin elicits extrarenal responses (see above) in patients with autosomal-recessive NDI, patients with X-linked NDI lack extrarenal responses to desmopressin (Bichet et al., 1988; Knoers and Monnens, 1991; Deen et al., 1994)

Identification of the molecular defects underlying congenital NDI is of immediate clinical significance, allowing diagnosis by gene analysis (see Figure 4). Gene analysis should be performed in newborns with a family history for NDI and patients of all age groups with a firm diagnosis of congenital NDI, with or without a family history. It may also be considered in babies presenting with continuing fever of unknown origin, vomiting, constantly low urine osmolality, and failure to thrive. Mutation analysis of NDI genes in babies is of particular value since the key symptoms of NDI, polyuria and polydipsia, are frequently not observed. As a consequence, patients of this age often remain undiagnosed and may suffer irreversible damage due to severe dehydration (see above). Gene analysis is also important for identification of nonobligatory female carriers of families with X-linked NDI. Most females that are heterozygous for a mutation in the V2 receptor do not present with clinical symptoms; few are severely affected (Oksche et al., 1994; van Lieburg et al., 1995).

All complications of congenital NDI are prevented by an adequate water intake. Thus patients should be provided with unrestricted amounts of water from birth to ensure normal development. The use of diuretics (thiazides) or indometacin may reduce urinary output. This advantagous effect has to be weighed against the side effects of these drugs (thiazides: electrolyte disturbances; indomethacin: reduction of the glomerular filtration rate and gastrointestinal symptoms). With the identification of the genes responsible for congenital NDI, a causative treatment based on gene transfer has become possible. Two prerequisites crucial for gene therapy seem to be fullfiled in both forms of congenital NDI: (1) The defect in the kidney appears to be restricted to water reabsorption with no other functional or histological defects. Unlike central diabetes insipidus (Schmale et al., 1993), retinitis pigmentosa (see above) and many other diseases, a deterioration of kidney function due to progessive structural changes is not observed. Thus the organ integrity seems to be preserved. (2) Recent experiments with rats show that adenoviral-mediated gene transfer to the tubular system of the kidney can be achieved either by selective perfusion of the renal artery or by retrograde infusion through a catheter placed into the pelvic cavity (Moullier et al., 1994). Depending on the route, expression of the reporter gene (β-galactosidase) is observed in proximal tubule cells (kidney perfusion via renal artery) or tubular cells of the papilla and medulla (retrograde infusion).

The mutations in the V2 receptor associated with X-linked NDI were the first naturally occurring mutations found in the very large group of G protein-coupled hormone receptors. Within the last few years however, a number of diseases were shown to be caused by mutations in genes encoding G protein-coupled receptors. In addition to retinitis pigmentosa and X-linked NDI, examples of such diseases/symptoms are stationary night blindness, color blindness/altered color perception, primary adrenocortical deficiency, hypocalciuric hypercalcemia/hyperparathyroidism, hypercalcemia/metaphyseal chondrodysplasia, hypocalcemia, male precocious puberty, male pseudohermaphroditism, hyperfunctioning thyroid adenoma, and Hirschsprung's disease (reviewed in Coughlin, 1994). Excepting somatic mutations in the thyroid-stimulating hormone receptor, mutations in G protein-coupled receptors are germ line mutations. Some mutations result in a hyperactive phenotype with a concomitant loss of receptor regulation by extracellular signals. Inactivating mutations may show a true functional defect, i.e., a decrease in the ability of the mutant protein to be activated by the extracellular signal or to transduce the signal to G proteins. There is, however, increasing evidence for the assumption that the main defect of many inactivating mutations is the reduced expression of mutant receptors on the cell surface. Here the loss of receptor function occurs regardless of the remaining biological activity of the individual protein. At present, very little is known about the cellular routing of G protein-coupled receptors. Progress in this field will be crucial for the understanding of the clinical phenotypes of receptor diseases on a molecular level and for the development of therapeutic strategies based on gene transfer.

ACKNOWLEDGMENTS

We thank John Dickson for critical reading of the manuscript. Our work reported herein was supported by grants of the Deutsche Forschungsgemeinschaft and a grant of the Thyssen-Stiftung to W.R.

REFERENCES

Agre, P., Brown, D., & Nielsen, S. (1995). Aquaporin water channels. Unanswered questions and unresolved controversies. Current Opinion Cell Biol. 7, 472-483.

Bai, L., Fushimi, K., Sasaki, S., & Marumo, F. (1996). Structure of aquaporin-2 vasopressin water channel. J. Biol. Chem. 271, 5171-5176.

Bichet, D.G., Arthus, M.F., Lonergan, M., Balfe, W., Skorecki, K., Nivet, H., Robertson, G., Oksche, A., Rosenthal, W., Fujiwara, M., Morgan, K., & Sasaki, S. (1995). Autosomal dominant and autosomal recessive nephrogenic diabetes insipidus: novel mutations in the AQP2 gene. J. Am. Soc. Nephrol. 6, 717.

Bichet, D.G., Arthus, M.-F., Lonergan, M., Hendy, G.N., Paradis, A.J., Fujiwara, T.M., Morgan, K., Gregory, M.C., Rosenthal, W., Antaramian, A., Didwania, A., & Birnbaumer, M. (1993). X-linked nephrogenic diabetes insipidus in North America and the Hopewell Hypothesis. J. Clin. Invest. 92, 1262-1268.

Bichet, D.G., Birnbaumer, M., Lonergan, M., Arthus, M.-F., Rosenthal, W., Goodyer, P., Nivet, H., Benoit, S., Giampietro, P., Simonetti, S., Fish, A., Whitley, C.B., Jaeger, P., Gertner, J., New, M., DiBona, F.J., Kaplan, B.S., Robertson, G.L., Hendy, G.N.,

Fujiwara, T.M., & Morgan, K. (1994). Nature and recurrence of AVPR2 mutations in X-linked nephrogenic diabetes insipidus. Am. J. Hum. Genetics 55, 278-286.

Bichet, D.G., Hendy, G.N., Lonergan, M., Arthus, M.-F., Ligier, S., Pausova, Z., Kluge, R., Zingg, H., Saenger, P., Oppenheimer, E., Hirsch, D.J., Gilgenkranzt, S., Salles, J.-P., Oberle, I., Mandel, J.-L., Gregory, M.C., Fujiwara, M., Morgan, K., & Scriver, C.R. (1992).: X-linked nephrogenic diabetes insipidus: From the ship Hopewell to RFLP studies. Am. J. Hum. Genet. 51, 1089-1102.

Bichet, D.G., Razi, M., Longergan, M., Arthus, M.-F., Vassiliki, P., Kortas, C., & Barjon, J.-N. (1988). Hemodynamic and coagulation responses to 1-desamino[8-D-Arginine] vasopressin in patients with congenital nephrogenic diabetes insipidus. N. Engl. J. Med. 318, 881-887.

Birnbaumer, M., Gilbert, S., & Rosenthal, W. (1994). An extracellular congenital nephrogenic diabetes insipidus mutation of the vasopressin receptor reduces cell surface expression, affinity for ligand, and coupling to the G_s/adenylyl cyclase system. Mol. Endocrinol. 8, 886-894.

Birnbaumer, M., Seibold, A., Gilbert, S., Ishido, M., Barberis, B., Antaramian, A., Brabet, P., & Rosenthal, W. (1992). Molecular cloning of the receptor for human antidiuretic hormone. Nature 357, 333-335.

Bode, H.H., & Crawford, J.D. (1969). Nephrogenic diabetes insipidus in North America - the Hopewell hypothesis. N. Engl. J. Med. 280, 750-754.

Breyer, M.D., & Ando, Y. (1994). Hormonal signaling and regulation of salt and water transport in the collecting duct. Annu. Rev. Physiol. 56, 711-739

Cannon, J.F. (1955). Progress in internal medicine: Diabetes insipidus. Arch. Int. Med. 98, 215-272.

Chini, B., Mouillac, B., Ala, Y., Balestre, M.-N., Trumpp-Kallmeyer, S., Hoflack, J., Elands, J., Hibert, M., Manning, M., Jard, S., & Barberis, C. (1995). Tyr115 is the key residue for determining agonist selectivity in the V1a vasopressin receptor. EMBO J. 14, 2176-2182.

Colley, N.J., Cassill, J.A., Baker, E.K., & Zuker, C.S. (1995). Defective intracellular transport is the molecular basis of rhodopsin-dependent dominant retinal degeneration. Proc. Natl. Acad. Sci. USA 92, 3070-3074.

Coughlin, S.R. (1994). Expanding horizons for receptors coupled to G-proteins: diversity and diesease. Current Biology 6, 191-197.

Deen, P.M.T., Croes, H., van Aubel, R.A.M.H., Ginsel, L.A., & van Os, C.H. (1995). Water channels encoded by mutant aquaporin-2 genes in nephrogenic diabetes insipidus are impaired in their cellular routing. J. Clin. Invest. 95, 2291-2296.

Deen, P.M.T., Verdijk, M.A.J., Knoers, N.V.A.M., Wieringa, B., Monnens, L.A.H., van Os, C.H., & van Oost, B.A. (1994). Requirement of human renal water channel aquaporin-2 for vasopressin-dependent concentration of urine. Science 264, 92-95.

Ecelbarger, C., Terris, J., Frindt, G., Echevarria, M., Marples, D., Nielsen, S., & Knepper, M.A. (1995). Aquaporin-3 water channel localization and regulation in rat kidney. Am. J. Physiol. 1995 269, F663-672.

Frigeri, A., Gropper, M.A., Kawashima, K., Umenishi, F., Brown, D., & Verkman, A.S. (1995). Localization of MIWC and GLIP water channel homologs in neuromuscular epithelia and glandular tissues. J. Cell Sci. 108, 2993-3002.

Fujiwara, T.M., Morgan, K., & Bichet, D.G. (1995). Molecular Biology of diabetes insipidus. Annu. Rev. Med. 46, 331-343.

Fushimi, K., Uchida, S., Hara, Y., Hirata, Y., Marumo, & Fk., Sasaki, S. (1993). Cloning and expression of apical membrane water channel of rat kidney collecting tubule. Nature 361, 549-552.

Handler, J.S. (1988). Antidiuretic hormone moves membranes. Am. J. Physiol. 255, F375-F382

Hays, R.M., Franki, N., Simon, H., & Gao, Y. (1994). Antidiuretic hormone and exocytosis: lessons from neurosecretion. Am. J. Physiol. 267, C1507-C1524.

Holtzman, E.J., Kolakowski, L.F., Jr., O'Brien, D., Crawford, J.D., & Ausiello, D.A. (1993). A null mutation in the vasopressin V2 receptor gene (AVPR2) associated with nephrogenic diabetes insipidus in the Hopewell kindred. Hum. Mol. Genet. 2, 1201-1204.

Jans, D.A., van Oost, B.A., Ropers, H.H., & Fahrenholz, F. (1990). Derivatives of somatic cell hybrids which carry the human gene locus for nephrogenic diabetes insipidus (NDI) express functional vasopressin renal V_2-type receptors. J. Biol. Chem. 265, 15379-15382.

Jo, I., Harris, H.W., Amedt-Raduege, A.M., Majweski, R.R., & Hammond, T.G. (1995). Rat kidney papilla contains abundant synaptobrevin protein that participates in the fusion of antidiuretic hormone (ADH) water channel-containing endosomes in vitro. Proc. Natl. Acad. Sci. USA 92, 1876-1880.

Jung, J.S., Preston, G.M., Smith, B.L., Guggino, W.B., & Agre, P. (1994). Molecular structure of the water channel through aquaporin CHIP. The hourglass model. J. Biol. Chem. 269, 14648-14654.

Katsura, T., Verbavatz, J.-M., Farinas, J., Ma, T., Ausiello, D.A., Verkman, A.S., & Brown, D. (1995). Constitutive and regulated membrane expression of aquaporin 1 and aquaporin 2 water channels in stably transfected LLC-PK$_1$ epithelial cells. Cell Biol. 92, 7212-7216.

Knepper, M.A. (1994). The aquaporin family of molecular water channels. Proc. Natl. Acad. Sci. USA 91, 6255-6258.

Knoers, N., & Monnens, L.A.H. (1991). A variant of nephrogenic diabetes insipidus: V2 receptor abnormality restricted to the kidney. Eur. J. Pediatr. 150, 370-373.

Knoers, N, van der Heyden, H., van der Oost, B.A., Monnens, L., Willems, J., & Ropers, H.H. (1988). Nephrogenic diabetes insipidus: close linkage with markers from the distal long arm of the human X chromosome. Hum. Genet. 30, 31-38.

Knoers, N.V.A.M., van den Ouweland, A.M.W., Verdijk, M., Monnens, L.A.H., & van Oost, B.A. (1994). Inheritance of mutations in the V_2 receptor gene in thirteen families with nephrogenic diabetes insipidus. Kidney Internat. 46, 170-176.

Kojro, E., Eich, P., Gimpl, G., & Fahrenholz, F. (1993). Direct identification of an extracellular agonist binding site in the renal V_2 vasopressin receptor. Biochemistry 32, 13537-13544.

Kuwahara, M., Fushimi, K., Terada, Y., Bai, L., Marumo, F., & Sasaki, S. (1995). cAMP-dependent phosphorylation stimulates water permeability of aquaporin-collecting duct water channel protein expressed in *Xenopus* oocytes. J. Biol. Chem. 270, 10384-10387.

Lande, M.B., Jo., I., Zeidel, M.L., Somers, M., & Harris, H.W. (1996). Phosphorylation of aquaporin-2 does not alter the membrane water permeability of rat papillary water channel-containing vesicles. J. Biol. Chem. 271, 5552-5557.

Liebenhoff, U., & Rosenthal, W. (1995). Idendification of Rab3-, Rab5a- and synaptobrevin-like proteins in a preparation of rat kidney vesicles containing the vasopressin-regulated water channel. FEBS Letters. 365, 209-213.

Lolait, S.J., O'Carroll, A.-M., McBride, O.W., Konig, M., Morel, A., & Brownstein, M.J. (1992). Cloning and characterization of a vasopressin V2 receptor and possible link to nephrogenic diabetes insipidus. Nature 357, 336-339.

Moullier, P., Friedlander, G., Calise, D., Ronco, P., Perricaudet, M., & Ferry, N. (1994). Adenoviral-mediated gene transfer to renal tubular cells *in vivo*. Kidney Internat. 45, 1220-1225.

Nielsen, S., Chou, C.L., Marples, D., Christensen, E.I., Kishore, B.K., & Knepper, M.A. (1995a). Vasopressin increases water permeability of kidney collecting duct by inducing translocation of aquaporin-CD water channels to plasma membrane. Proc. Natl. Acad. Sci. USA 92, 1013-1017.

Nielsen, S., DiGiovanni, S.R., Christensen, E.I., Knepper, M.A., & Harris, H.W. (1993a). Cellular and subcellular immunolocalization of vasopressin-regulated water channel in rat kidney. Proc. Natl. Acad. Sci. USA 90, 11663-11667.

Nielsen, S., Marples, D., Birn, H., Mohtashami, M., Dalby, N.O., Trimble, W., & Knepper, M. (1995b). Expression of VAMP2-like protein in kidney collecting duct intracellular vesicles. J. Clin. Invest. 96, 1834-1844.

Nielsen, S., Smith, B.L., Christensen, E.I., Knepper, M.A., & Agre, P. (1993b). CHIP28 water channels are localized in constitutively water-permeable segments of the nephron. J. Cell Biol. 120, 371-383.

Oksche, A., Dickson, J., Schülein, R., Seyberth, H.W., Müller, M., Rascher, W., Birnbaumer, M., & Rosenthal, W. (1994). Two novel mutations in the vasopressin V2 receptor gene in patients with congenital nephrogenic diabetes insipidus, Biochem. Biophys. Res. Comm. 205, 552-557.

Oksche, A., Möller, A., Dickson, J., Rosendahl, W., Rascher, W., Bichet, D.G., & Rosenthal, W. (1996a). Two novel mutations in the aquaporin-2 and the vasopressin V2 receptor genes in patients with congenital nephrogenic diabetes insipidus. Human Genet. 98, 587-589.

Oksche, A., Schülein, R., Rutz, C., Liebenhoff, U., Dickson, J., Müller, H., Birnbaumer, M., & Rosenthal, W. (1996b). Vasopressin V2 receptor mutants causing X-linked nephrogenic diabetes insipidus: analysis of expression, processing and function. Mol. Pharmacol. 50, 820-828.

Pan, Y., Metzenberg, A., Das, S., Jing, B., & Gitschier, J. (1992). Mutations in the V2 vasopressin receptor gene are associated with X-linked nephrogenic diabetes insipidus. Nature Genet. 2, 103-106.

Pan, Y., Wilson, P., & Gitschier, J. (1994). The effect of eight V2 vasopressin receptor mutations on stimulation of adenylyl cyclase and binding to vasopressin. J. Biol. Chem. 269, 31933-31937.

Preston, G.M., & Agre, P. (1991). Molecular cloning of the red cell integral membrane protein of M_r 28,000: a member of an ancient channel family. Proc. Natl. Acad. Sci. USA 88, 11110-11114.

Preston, G.M., Smith, B.L., Zeidel, M.L., Moulds, J.J., & Agre, P. (1994). Mutations in aquaporin-1 in phenotypically normal humans without functional chip water channels. Science 265, 1585-1587.

Raina, S., Preston, G.M., Guggino, W.B., & Agre, P. (1995). Molecular cloning and characterization of an aquaporin cDNA from salivary, lacrimal, and respiratory tissues. J. Biol. Chem. 270, 1908-1912

Rittig, S., Robertson, G.L., Siggaard, C., Kovács, L., Gregersen, N., Nyborg, J., & Pedersen, E.B. (1996). Identification of 13 new mutations in the vasopressin-neurophysin II gene in 17 kindreds with familial autosomal dominant neurohypophyseal diabetes insipidus. Am. J. Hum. Genet. 58, 107-117.

Rosenthal, W., Antaramian, A., Gilbert, S., & Birnbaumer, M. (1993). Nephrogenic diabetes insipidus. J. Biol. Chem. 268, 13030-13033.

Rosenthal, W., Seibold, A., Antaramian, A., Lonergan, M., Arthus, M.-F., Hendy, G., Birnbaumer, M., & Bichet, D. (1992). Molecular identification of the gene responsible for congenital nephrogenic diabetes insipidus. Nature 359, 233-235.

Ruiz-Opazo, N., Akimoto, K., & Herrera, V.L.M. (1995). Identification of a novel dual angiotensin II/vasopressin receptor on the basis of molecular recognition theory. Nature Med. 1, 1074-1081.

Sabolic, I., Valenti, G., Verbavatz, J.M., van Hoek, A.N., Verkman, A.S., Ausiello, D.A., & Brown, D. (1992). Localization of the CHIP28 water channel in rat kidney. Am. J. Physiol. 263, C1225-C1233.

Sabolic, I., Katsura, T., Verbavatz, J.M., & Brown, D. (1995). The AQP2 water channel: effect of vasopressin treatment, microtubule disruption, and distribution in neonatal rats. J. Membr. Biol. 143, 165-175.

Sasaki, S., Fushimi, K., Saito, H., Saito, F., Uchida, S., Ishibashi, K., Kuwahara, M., Ikeuchi, T., Inui, K., Nakajima, K., Watanabe, T.X., & Marumo, F. (1994). Cloning, characterization, and chromosomal mapping of human aquaporin of collecting duct. J. Clin. Invest. 93, 1250-1256.

Schmale, H., Bahnsen, U., & Richter, D. (1993). Structure and expression of the vasopressin precursor gene in central diabetes insipidus. Ann. New York Acad. Sci. 689, 74-82.

Schülein, R., Rutz, C., & Rosenthal, W. (1996). Membrane targeting and determination of transmembrane topology of the human vasopressin V2 receptor. J. Biol. Chem. In revision.

Seibold, A., Brabet, P., Rosenthal, W., & Birnbaumer, B. (1992). Structure and chromosomal localization of the human antidiuretic hormone receptor gene. Am. J. Hum. Genet. 51, 1078-1083.

Souied E., Gerber S., Rozet, J.-M., Bonneau, D., Dufier, J-L., Ghazi, I., Philip, N., Soubrane, G., Coscas, G., Munnich, A., & Kaplan, J. (1994). Five novel missense mutations of the rhodopsin gene in autosomal dominant retinitis pigmentosa. Hum. Mol. Gen. 3, 1433-1434.

Sugimoto, T., Saito, M., Mochizuki, S., Watanabe, Y., Hashimoto, S., & Kawashima, H. (1994). Molecular cloning and functional expression of a cDNA encoding the human V_{1b} vasopressin receptor. J. Biol. Chem. 269, 27088-27092.

Sung, C.-H., Davenport, C.M., & Nathans, J. (1993). Rhodopsin mutations responsible for autosomal dominant retinitis pigmentosa. J. Biol. Chem. 268, 26645-26649.

Sung, C.-H., Makino, C., Baylor, D., & Nathans, J. (1994). A rhodopsin gene mutation responsible for autosomal dominant retinitis pigmentosa results in a protein that is defective in localization to the photoreceptor outer segment. J. Neurosci. 14, 5818-5833.

Tajima, T., Nakae, J., Takekoshi, Y., Takahashi, Y., Yuri, K., Nagashima, T., & Fujieda, K. (1996). Three novel AVPR2 mutations in three Japanese families with X-linked nephrogenic diabetes insipidus. Pediatric Res. 39, 522-526

Terris, J., Ecelbarger, C., Marples, D., Knepper, M.A., & Nielsen, S. (1995). Distribution of aquaporin-4 water channel expression within rat kidney. Am. J. Physiol. 269, F775-F785.

Thibonnier, M., Auzan, C., Madhun, Z., Wilkins, P., Berti-Mattera, L., & Clauser, E. (1994). Molecular cloning, sequencing, and functional expression of a cDNA encoding the human V_{1a} vasopressin receptor. J. Biol. Chem. 269, 3304-3310.

Tsukaguchi, H., Matsubara, H., & Inada, M. (1995a). Expression studies of two vasopressin V2 receptor gene mutations, R202C and 804insG, in nephrogenic diabetes insipidus. Kidney Internat. 48, 554-562.

Tsukaguchi, H., Matsubara, H., Mori, Y., Yoshimasa, Y., Yoshimasa, T., Nakao, K., & Inada, M. (1995b). Two vasopressin type 2 receptor gene mutations R143P and Δ V278 in patients with nephrogenic diabetes insipidus impair ligand binding of the receptor. Biochem. Biophys. Res. Commun. 211, 967-977.

Tsukaguchi, H., Matsubara, H., Taketani, S., Mori, Y., Seido, T., & Inada, M. (1995c). Binding-, intracellular transport-, and biosynthesis-defective mutants of the vasopressin type 2 receptor in patients with X-linked nephrogenic diabetes insipidus. J. Clin. Invest. 96, 2043-2050.

Uchida, S., Sasaki, S., Fushimi, K., & Marumo F. (1994). Isolation of human aquaporin-CD gene. J. Biol. Chem. 269, 23451-23455.

Vaithinathan, R., Berson, E.L., & Dryja, T.P. (1994). Further screening of the rhodopsin gene in patients with autosomal dominant retinitis pigmentosa. Genomics 21, 461-463.

van den Ouweland, A.M.W., Dreesen, J.C.F.M., Verdijk, M., Knoers, N.V.A.M., Monnens, L.A.H., Rocchi, M., & van Oost, B.A. (1992). Mutations in the vasopressin type 2 receptor gene (AVPR2) associated with nephrogenic diabetes insipidus. Nature Genet. 2, 99-102.

van Lieburg, A.F., Verdijk, M.A.J., Knoers, N.V.A.M., van Essen, A.J., Proesmans, W., Mallmann, R., Monnens, L.A.H., van Oost, B.A., van Os, C.H., & Deen, P.M.T. (1994). Patients with autosomal nephrogenic diabetes insipidus homozygous for mutations in the aquaporin 2 water-channel gene. Am. J. Hum. Genet. 55, 648-652.

van Lieburg, A.F., Verdijk, M.A.J., Schoute, F., Ligtenberg, M.J.L., van Oost, B.A., Waldhauser, F., Dobner, M., Monnens, L.A.H., & Knoers, N.V.A.M. (1995). Clinical phenotype of nephrogenic diabetes insipidus in females heterozygous for a vasopressin type 2 receptor mutation. Human. Genet. 96, 70-78.

Valenti, G., Frigeri, A., Ronco, P., & Svelto, M. (1995). A Human Collecting Duct Cell Line Expresses Members of the Aquaporin Family. Aquaporins and Epithelial Water Transport. International Symposium Manchester, abstract volume, p. 25.

Wade, J.B., Stetson, D.L., & Lewis, S.A. (1981). ADH action: evidence for a membrane shuttle mechanism. Ann. NY Acad. Sci. 372, 106-117.

Wildin, R.S., Antush, M.J., Bennett, R.L., Schoof, J.M., & Scott, C.R. (1994). Heterogeneous AVPR2 gene mutations in congenital nephrogenic diabetes insipidus. Am. J. Hum. Genet. 55, 266-277.

Chapter 8

Mechanisms of Radiation-Induced Carcinogenesis: The Thyroid Model

YURI E. NIKIFOROV and JAMES A. FAGIN

Advances in Molecular and Cellular Endocrinology
Volume 2, pages 169-196.
Copyright © 1998 by JAI Press Inc.
All rights of reproduction in any form reserved.
ISBN: 0-7623-0292-5

INTRODUCTION: THYROID CANCER AS A MODEL OF RADIATION-INDUCED TUMORS

Ionizing radiation exposure is a well-established risk factor for a number of human solid neoplasms and hematologic malignancies. The common known sources of ionizing radiation responsible for health effects are: 1) medical irradiation (therapeutic or diagnostic), which is more often external and well-documented in terms of doses and fields of exposure; and 2) accidental environmental irradiation, which has a much less predictable range of doses, target tissues, and eventually health consequences. Thyroid cancer represents a striking example of tumors which have a well-documented association with both these types of radiation exposure.

The association between radiation to the thyroid gland and thyroid cancer development was first proposed in 1950 in children who received X-ray therapy for an enlarged thymus (Duffy and Fitzgerald, 1950). Since then, numerous reports have documented increased incidence of thyroid neoplasms in patients with a history of prior irradiation for different benign conditions of the head, neck, and thorax (Winship and Rosvoll, 1970). Most of these tumors were diagnosed in the United States between the 1940s and 1960s. After 1965, when the use of radiation therapy for benign conditions was almost completely abandoned, the incidence of childhood thyroid malignancy decreased dramatically. Since then, radiation therapy for a variety of malignancies continues to be a source of radiation-associated thyroid cancer.

Environmental irradiation is also a causal factor for thyroid cancer, as suggested after long-term follow-up of survivors exposed to gamma and neutron radiation due to the atomic bomb explosions in Japan in 1945. The increased risk for thyroid cancer was higher in individuals exposed at a younger age (Prentice et al., 1982; Ezaki et al., 1991). The incidence of thyroid cancer was also increased among residents of the Marshall Islands exposed to fallout radiation after detonation of a thermonuclear device on the Bikini atoll in 1954 (Conard, 1984). The exposure to the thyroid gland resulted from internal irradiation from absorbed short-lived radioiodines and, to a lesser extent, penetrating gamma radiation. Appearance of thyroid carcinomas was noted 10 years after the exposure, with seven papillary, one follicular, and seven occult thyroid cancers diagnosed among the 250 exposed individuals during 34 years of careful monitoring (Cronkite et al., 1995).

Appearance of thyroid cancer in children after environmental exposure to radiation was noted soon after the accident at the Chernobyl nuclear power plant in the former USSR in April 1986, where millions of curies of short-lived radioiodine isotopes were released in the fallout and became a source of internal exposure to the thyroid gland. An increased incidence of thyroid cancer in children from the most contaminated areas of Belarus was noted as early as four years after the accident (Kazakov et al., 1992). In 1991–1992, the incidence of childhood thyroid cancer in Belarus reached a level that was 60-fold greater than prior to the disaster (Nikiforov and Gnepp, 1994).

Common features characteristic of all radiation-induced thyroid cancers are: a latent period of 4–45 years after exposure, linear dose-response relationship, strong inverse correlation with age at the time of exposure, and a predominantly papillary type of thyroid cancer (Nikiforov and Fagin, 1997).

In spite of the fact that association between radiation exposure and thyroid cancer has been established for several decades, the precise molecular mechanisms of radiation-induced carcinogenesis in the thyroid gland remain unclear. In this chapter, we will review some of the current information on the mechanisms of radiation-induced DNA damage and mutagenesis. We will also discuss our present knowledge of the epidemiological and pathological characteristics and genetic events associated with the post-Chernobyl pediatric thyroid neoplasms, and make some predictions about the possible mechanisms responsible for radiation-induced carcinogenesis in the thyroid gland.

MOLECULAR MECHANISMS ASSOCIATED WITH RADIATION TUMORIGENESIS

Immediate Effects of Radiation on Genome

The biological effects of ionizing radiation, including cell killing, mutagenesis, and transformation, are generally considered to originate from direct and indirect damage to DNA. The direct damage comes from energy directly deposited in DNA molecules, while indirect damage arises following attack on the DNA by reactive particles produced by ionizations of other molecules, primarily by free radicals originated from water (Ward, 1988). DNA damage from ionizing radiation results in single-strand breaks (SSBs), double-strand breaks (DSBs), and a wide range of altered bases (Ward, 1988; Goodhead, 1989). A radiation dose of 2 Gy produces approximately 2000 SSBs, 80 DSBs, and 6000 damaged bases per cell (Ward, 1994). The initial damage produced by either direct ionization or by free radical attack occurs randomly throughout the genome, as demonstrated by tracking the fate of selected chromosomes from irradiated cells with fluorescent *in-situ* hybridization (FISH) after fusion with non-irradiated HeLa cells to obtain chromosome condensation (Kovacs et al., 1994).

Active enzymatic mechanisms exist for the repair of DNA damage. Misrepair of DNA DSBs is considered to be the major source of mutation formation, since in the absence of concomitant DNA replication, individual SSBs are not thought to contribute significantly to this process. This is because SSBs are repaired rapidly and accurately (with a half-life of four minutes, even at room temperature (Blakely et al., 1982)), using the complementary undamaged strand as a template, so that the probability of their misrepairing is potentially very low. Unrepaired DSBs are lethal for the cell, while misrepaired DSBs introduced into the genome can result in the formation of chromosome aberrations and other mutations. In cultured human fibroblasts irradiated with 6 Gy, about 85% of the initial brakes have been restituted after 24 hours. Of the remaining 15%, approximately half were involved in exchanges, and the other half remained unrejoined (Kovacs et al., 1994). Days later, severe chromosomal alterations such as dicentrics were gradually lost from the irradiated populations as the cells containing them died, but the incidence of translocations remained constant.

The majority of data on the spectrum of mutations induced by radiation were obtained on model systems which work by selecting for the loss of function of a gene product that confers cellular sensitivity to a drug, whereas non-mutated cells are killed. Such systems include loss of function of the enzyme hypoxanthine-guanine phosphoribosyltransferase (HPRT), which renders cells resistant to the drug 6-thioguanine, thymidine kinase (TK), giving resistance to trifluorothymidine, or adenine phosphoribosyltransferase (APRT), giving resistance to 8-azaadenine. After 5 Gy γ-irradiation of cultured hamster cells, numerous large chromosomal alterations and point mutations were found in the APRT and HPRT genes (Grosovsky et al., 1986; 1988; Thacker, 1986). In different experiments, 16–70% of radiation-induced mutations were large chromosomal alterations (>50 bp), which included total or partial gene deletions and chromosomal rearrangements. The others were classified as point mutations,

being below the resolution of Southern blotting, and included small deletions, frameshifts, and a wide variety of base substitutions. The types of
mutations were consistent in the various loci examined. The proportion of
large chromosomal alterations among radiation-induced mutations was always higher than among spontaneously occurring mutations in the same
genes, since the latter were almost exclusively point mutations. As op-
posed to spontaneous mutations, radiation-induced mutations in these
model systems lacked obvious hot spots and were randomly distributed
through the gene (Grosovsky et al., 1988).

Recently, human cell models were also employed to study radiation
mutagenesis. The spectrum of mutations induced by 2 Gy irradiation in
HPRT and TK genes from cultured human lymphoblasts includes large
scale alterations, primarily deletions, in 67–68%, and point mutations in
32–33% (Nelson et al., 1994; Giver et al., 1995). The types of point mutations were very diverse and included all classes of transitions and transversions. However, transitions were relatively less common in
X-ray-induced than in spontaneously occurring mutants (Giver et al.,
1995). Tandem base substitutions, frameshifts, small deletions, and a de-
letion/insertion compound mutation were also observed. In general, insertion mutations were relatively infrequent in radiation-induced HPRT⁻
and TK⁻ mutants. These recent studies on human cells underlined the predominance of large-scale chromosomal abnormalities. However, this
methodology may underestimate the actual prevalence of point mutations after irradiation, as those that lead to base substitutions that do not
impair or abolish the function of the respective enzyme will not be selected for in this assay. In some of the human cell models radiation-
induced mutations were found to cluster at particular sites of the gene
(Giver et al., 1995). Radiation-induced deletions and rearrangements
were commonly found in spots surrounded by direct DNA repeats.

Possible Mechanisms of Radiation-Induced Transformation

Although there is considerable data on the mechanisms of radiation
DNA damage and mutagenesis on the one side, and the characterization
of radiation-induced tumors on the other, there is an unbridged gap between these two end points, resulting in uncertainty as to the precise molecular mechanisms of radiation carcinogenesis.

Two possible mechanisms were proposed by John Little in his Failla
Memorial lecture (Little, 1994). Radiation may lead directly to onco-

genic mutation(s) as a result of misrepaired DNA damage in a cell, and all progeny of this initial cell would carry the same mutation and generate the malignant clone (Figure 1, top). Alternatively, radiation might induce a process leading to a type of genomic instability that is transmitted through subsequent cell divisions, and that increases the probability of later occurrence of transforming mutations (Figure 1, bottom). This latter sequence of events was named the delayed or indirect mechanism of radiation tumorigenesis.

These hypotheses are illustrated by the classic transformation experiments performed by Little and his colleagues. Irradiated cells were seeded at low density, and 4–5 weeks after reaching confluency, foci of transformation appeared overlaying the normal monolayer, suggesting that irradiation directly initiated malignant transformation in an occasional cell (Terzaghi and Little, 1976). However, subsequent experiments by this group challenged this notion, by demonstrating that the appearance of transformed foci was not proportional to the number of cells seeded initially, suggesting that radiation first induces genomic in-

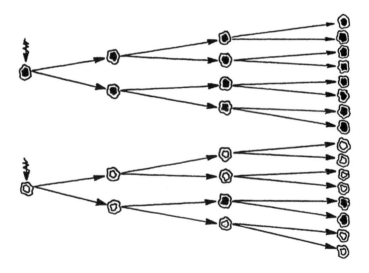

Figure 1.   Two possible mechanisms of radiation-induced transformation. Top: radiation-induced DNA damage and misrepair leads directly to transforming mutation(s) in one cell, giving rise to a malignant clone. Bottom: radiation induces genomic instability that is transmitted through subsequent cell divisions, increasing the probability of later occurrence of transforming mutations. Modified from Little (1994).

stability as a high frequent event, and that secondary transforming mutations may result as low frequency events (Little, 1994).

The concept that radiation may directly lead to transforming genetic mutations is also supported by the fact that the two most common types of human malignancies associated with radiation, leukemias and papillary thyroid carcinomas, have chromosomal rearrangements as important oncogenic events. As discussed above, large-scale genomic abnormalities, including rearrangements, are the most common outcome of misrepaired DNA damage after cell irradiation *in vitro*. Moreover, high-dose γ-radiation has been shown to induce *ret* gene rearrangements in human thyroid cells *in vitro* (Ito et al., 1993).

On the other hand, there is good experimental evidence that radiation may not result in the formation of transforming genes as a direct result of its DNA damaging properties, but rather that it predisposes to delayed genetic events through the induction of an unstable genome in a progeny of irradiated cells (Kronenberg, 1994; Kadhim et al., 1995). Thus, DNA fingerprinting analysis of radiation-induced rat tumors revealed a high frequency of band shifts, gain or loss of amplified sequences, deletions, and appearance of extra bands in progressive biopsies, which was interpreted as evidence for genomic destabilization (Felber et al., 1994). It has also been shown that mutations scored soon after irradiation differ significantly in their spectrum from late-arising mutations (Little, 1994). In the former case, 73% of mutations were deletions and 27% point mutations, while among late-arising mutations the ratio was exactly opposite.

It has been hypothesized that radiation may lead to delayed accumulation of point mutations through defects in mismatch repair, as described for hereditary nonpolyposis colorectal cancer cells (Karran, 1995). This DNA mismatch repair deficit, due to loss-of-function of one of the DNA repair enzymes (hMSH2, hMLH1, hPMS1, or PMS2), manifests as accumulation of numerous mutations affecting a number of simple tandem DNA repeats. Microsatellite instability has been found with high prevalence in inherited cancer syndromes such as hereditary non-polyposis colorectal cancer, and has also been observed in sporadic colorectal and endometrial carcinomas, indicating that critical elements of the DNA mismatch-repair machinery can be damaged as part of acquired, somatic events (Shibata et al., 1994).

Another type of genomic instability, minisatellite instability, has also been suggested to play a role in the generation of radiation-induced tumors. Minisatellite instability manifests as changes in the number of longer tandem DNA repeats. The mechanism and the precise DNA enzymatic repair mechanism responsible for this phenomenon is unknown. In mice, paternal irradiation results in an extremely high frequency of new minisatellite repeat polymorphisms in a selected locus in the offspring (Sadamoto et al., 1994), supporting the concept that radiation induces a "recombination-competent state" (Kronenberg, 1994). Minisatellite instability was also reported with high frequency in the offspring of irradiated mice (Dubrova et al., 1993), in humans (Dubrova et al., 1996), and in xenografts of X-ray transformed mouse cells growing *in vivo* (Paquette and Little, 1994).

Modulation of Radiation Mutagenesis and Transformation: Postirradiation Recovery and Cellular Response

The dose-rate effects for mutagenesis and eventually for transformation in cellular systems are believed to be strongly influenced by postirradiation repair and recovery processes. Their importance is best demonstrated by the existence of rare human disorders, such as ataxia-telangiectasia (AT), xeroderma pigmentosum, trichothiodystrophy and others, which are characterized by defective DNA repair and increased sensitivity to ionizing and/or ultraviolet (UV) radiation. Among these syndromes, AT and Nijmegen Breakage Syndrome are the best examples of disorders where affected individuals are also predisposed to cancer development (Murnane and Kapp, 1993). In order to recover efficiently, the affected cell should first undergo a reversible cell cycle arrest to allow DNA repair to take place, for which the functional DNA repair systems are critical.

Cell Cycle Control

Ionizing radiation induces the expression of a large number of early-response genes associated with many different cellular processes including signal transduction, intercellular signaling, growth control, DNA repair, and other potentially protective responses (Fornace, 1992). Ionizing radiation has been known for many years to induce a transient arrest in G_1/S and G_2 phases of the cell cycle in a wide range of eukaryotic cells. These delays presumably permit repair of DNA damage before starting replicative DNA synthesis or beginning mitosis. Wild-type p53 protein plays a critical role in G_1 cell cycle arrest after radiation exposure (Kastan et al., 1991; Kuebritz et al., 1992). The level of p53 protein rises 1–2 hours following γ irradiation and remains high for up to 72 hours. This rise is associated with G_1 arrest in cells expressing wild-type, but not mutant, p53 protein. This concept, however, has been challenged recently by evidence that tumor cell lines expressing wild-type p53 fail to arrest in the G_1 checkpoint of the first cell cycle after radiation (Nagasawa et al., 1995).

p53 acts as a transcriptional regulator and activates WAF1/CIP1 gene expression following irradiation. The former gene encodes a p21 protein, an inhibitor of multiple cyclin-dependent kinase/cyclin complexes, that is thought to be immediately responsible for G_1 arrest. GADD45 is another gene whose expression following DNA damage is dependent upon wild-type p53 (Zhan et al., 1994). GADD45 is a member of the GADD family of genes, that are inducible by different DNA damaging agents. GADD45 is the only one of these that is induced by g-radiation. The above pathway was also tested after radiation damage of human thyroid cells (Namba et al., 1995). Although the p53-WAF1/CIP1 pathway was found to be responsible for radiation-induced G_1 arrest in thyrocytes, this was not associated with induction of GADD45, at least after doses of 0.5–10 Gy.

As mentioned above, patients with AT display abnormal radiosensitivity as well as an increased rate of lymphoid cancer and possibly of breast cancer. Cultured cells derived from AT homozygotes show an abnormal response to ionizing radiation, such as increased cell death, radioresistant DNA synthesis following exposure to X-rays, and a higher level of spontaneous and radiation-induced chromosomal aberrations (Taylor et al., 1975). The increased *in vivo* and *in vitro* radiosensitivity in AT was found to be due to a deficit of postirradiation repair processes (Cox et al., 1981). In a model system where DSBs were produced by restriction enzyme cutting and then rejoined by human cell extracts, the cell extracts from patients with AT gave an elevated frequency of missrejoining, indicating loss of fidelity in repairing DNA breaks (North et al., 1990). Recently, the mammalian cell cycle G_1 checkpoint pathway which utilizes p53 and GADD45 has been shown to be defective in AT cells (Kastan et al., 1992). It appears that in unaffected cells, radiation induces expression of the AT gene, which in turn activates p53 expression followed by G_1 arrest—the checkpoint function lost in AT homozygotes.

The mechanisms responsible for the G_2 delay following irradiation, which is not p53 dependent, are not clear yet. Reduced expression of cyclin

B after irradiation may contribute to this, and the duration of the G_2 delay may also be influenced by the activation state of the cyclin B/p34^{cdc2} complex (Bernhard et al., 1995). In yeast, the product of the RAD9 gene is a critical factor in the G_2 arrest after g-irradiation (Weinert and Hartwell, 1988). Its possible homologues in mammalian cells have not been found.

DNA Repair

UV-induced DNA damage results in pyrimidine dimer formation that is primarily repaired by the nucleotide excision repair system. Other systems present in mammalian cells, such as recombinational, base excision, and post-replication repair, play a key role in the repair of DNA damage induced by ionizing radiation. UV-sensitive hamster cell mutants served as an important tool in the cloning of several human genes involved in the excision repair pathway. Because of this, several X-ray-sensitive rodent cell lines have now been identified, and are used to isolate genes involved in radiation DNA damage repair by using human cell extracts to complement the defect in radiation sensitive mutants (Collins, 1993). To date, at least 11 complementation groups have been identified, indicating the great complexity of the cellular response to ionizing radiation (Zdzienicka, 1995). Among them, five complementation groups are defective in DSBs rejoining and two groups are defective in SSBs repair. The human DSBs repair gene XRCC4 (X-ray cross-complementing gene 4), which complements XR-1 cells, has been mapped to chromosome 5 (Giaccia et al., 1990), and XRCC5, which complements XRS cells, to chromosome 2 (Chen et al., 1992). In addition, the XRCC1 and XRCC3 human genes involved in SSBs repair have been cloned by using X-ray-sensitive rodent mutants (Thompson et al., 1990; Tebbs et al., 1995). However, the precise function of these genes in SSBs repair remains unclear.

In each organism several mechanisms are present to repair DSBs. In addition to end-to-end rejoining, DSBs can be repaired by resection-annealing or by recombinational repair. Genes belonging to the RAD52 epistasis group are required for the recombinational repair of DSBs in yeast. Evolutionary conservation of DNA repair genes has enabled the cloning of human genes homologous to RAD52 (Muris et al., 1994), and another human gene homologous to RAD51 which is also involved in DSBs repair in yeast (Shinohara et al., 1993). However, the function of these recently cloned human homologues remains unknown.

RADIATION-INDUCED THYROID TUMORS AFTER CHERNOBYL

Post-Chernobyl Pediatric Thyroid Carcinomas as a Unique Model of Radiation-Induced Cancer in Humans

The Chernobyl accident took place in the former USSR on April 24, 1986, when an explosion at the fourth unit of the Chernobyl Nuclear Power Station destroyed the nuclear reactor and released millions of curies of radioactive materials into the atmosphere (Ilyin et al., 1990). During the initial period after the accident the most biologically significant isotopes released in the fallout were radioiodines, predominantly ^{131}I and other shorter-lived iodines (^{132}I, ^{133}I). The absorption of radioiodines from ingestion of contaminated food and water and through inhalation led to internal exposure to the thyroid gland, which was estimated to be 3–10 times higher in children than in adults. About 10% of the thyroid dose came from external irradiation. Due to the weather conditions immediately after the accident, about 70% of the most contaminated areas were located in the territory of the Republic of Belarus, especially in the Southern areas of Belarus, which lie immediately to the north of the Chernobyl plant. Current dose estimates suggest that thyroid doses in children under seven years of age from these areas were among the highest, and ranged between 15 and 100 cGy, with about 10% of children receiving more than 500 cGy (Becker et al., 1996). Other affected areas included parts of Northern Ukraine and Western Russia.

During the subsequent four years after the accident, substantial changes of thyroid function were detected in the exposed children, especially in those less than seven years of age (Astakhova et al., 1991; Polyanskaya et al., 1991). Beginning in 1990, a dramatic increase in the frequency of pediatric thyroid cancer was noted in Belarus. In 1986, there were two cases diagnosed in children less than 15 years of age; in 1987, four cases; in 1988, five cases; in 1989, six cases; in 1990, 23 cases, and in 1991, 55 cases (Kazakov et al., 1992).The histological examination of 104 thyroid glands by an international group of pathologists confirmed the diagnosis of malignancy in 102 cases (Baverstock et al., 1992). After 1991, the increase in frequency of childhood thyroid cancer continued, with 424 cases diagnosed in Belarus between 1986 and 1995 (International Conference "One decade after Chernobyl," 1996). In other affected regions the increased rate of thyroid cancer was noted after a longer latency, with 209 cases diagnosed in five Northern regions of the Ukraine, and 23 cases diagnosed in two Western regions of Russia in 1986–1994 (Stsjazhko et al., 1995).

The first reports on increased incidence of thyroid carcinoma in Belarussian children were met with some skepticism by the international community, partially because of the unprecedented short time after exposure, and the sharp and progressive increase observed. In addition, there were only indirect estimates of individual thyroid doses available for patients with thyroid carcinomas, since direct measurements were limited due to the large scale of the disaster, and the short half-life of radioiodines. Some suggested that the observed thyroid cancers may have been due to the heightened awareness and greater surveillance by health agencies after the nuclear accident. A decade after the accident, the evidence clearly indicates that the increased rate of pediatric thyroid cancer is real, and that this phenomenon is directly associated with the Chernobyl accident in 1986. The major pieces of evidence in support of these statements are as follows:

1. Sporadic thyroid carcinoma is a very rare disease in children all over the world. For example, in England and Wales the annual incidence of thyroid cancer has been estimated at 0.19 per million in children less than 15 years of age (De Keyser and Van Herle, 1985), and in Norway at 1.0 (Akslen et al., 1990). In the United States, for the period from 1973 to 1982, this rate was 1.8 per million per year (Young et al., 1986). These data are similar to the incidence of thyroid carcinoma in Belarus before 1986. According to the Belarussian Cancer Registry, nine thyroid carcinomas were diagnosed in patients younger than 15 years of age in 1976–1985, resulting in an annual incidence rate of 0.4 cases per million children per year (Okeanov and Averkin, 1991). By contrast, in 1991–1992, the total number of morphologically verified pediatric thyroid carcinomas was 116, with an incidence of 24.7 cases per million children per year, indicating a 62-fold increase in rate in 1991–1992 as compared to 1976–1985 (Nikiforov and Gnepp, 1994). The numbers calculated for the Gomel region, the most contaminated by the fall-out, are even more dramatic with a rate of 0.5 cases per million children per year in 1981–1985, and 96.4 in 1991–1994, resulting in a 193-fold increase (Stsjazhko et al., 1995).

2. If such high incidence were due to intense screening, it should result in the finding of a large number of early or occult tumors.

However, the opposite was true for post-Chernobyl thyroid cancers, since the latter were mostly diagnosed at advanced stages of disease. Thus, in a series of 84 tumors operated consecutively in 1991–1992, 88% of patients had tumor nodules greater than 1 cm in diameter, and an identical proportion of the tumors presented with lymph node metastases (Nikiforov and Gnepp, 1994). Only 7% of these neoplasms were occult, possibly detected because of the increased screening.

3. The incidence of pediatric thyroid carcinomas was proportional to the levels of radioiodine contamination of particular areas. Thus, the incidence of thyroid carcinoma in Belarus was 19 per million children in areas of ^{131}I contamination at levels of 5–10 Ku/km^2, 159 in areas with 40–50 Ku/km^2, 311 in areas with 50–100 Ku/km^2, and 976 in areas with >100 Ku/km^2 (Medical Consequences of the Catastrophe at the Chernobyl AES in Belarus, 1996). Among 333 children with thyroid carcinoma operated on in Belarus in 1986–1994, 54% resided at the time of the accident in the Gomel region, which is nearest to Chernobyl, whereas only 2% were from the Vitebsk region which has a comparable population, but no significant radiation fall-out (Medical Consequences of the Catastrophe at the Chernobyl AES in Belarus, 1996).

4. The thyroid doses reconstructed for 11 children with thyroid carcinoma ranged between 30 and 1,000 cGy, with a mean of 208±82 cGy (Medical Consequences of the Catastrophe at the Chernobyl AES in Belarus, 1996). Morphological examination of tumor tissues adjacent to tumor nodule revealed a high frequency of changes usually associated with a history of radiation exposure, including interstitial fibrosis, vascular abnormalities, and follicular atrophy (Nikiforov and Gnepp, 1994; Nikiforov et al., 1995). In addition, cytogenetic analysis of lymphocytes from children with thyroid carcinoma showed a significant increased incidence of other markers of radiation exposure, such as dicentric and ring chromosomes (Medical Consequences of the Catastrophe at the Chernobyl AES in Belarus, 1996).

5. The increased incidence of thyroid cancer after April 1986 had a characteristic age distribution of the affected children. Among 333 patients with thyroid carcinoma diagnosed in Belarussian children in 1986–1994, only seven were born after the Chernobyl

disaster (Stsjazhko et al., 1995). The average age of patients at clinical presentation has tended to increase with time, as it was 8.3±2.9 years in 1990, 9.0±2.9 in 1991, and 10.4±2.7 in 1992.

Based on this information, it is safe to conclude that the appearance of pediatric thyroid carcinomas in this population is directly associated with radiation exposure in April 1986. Other factors, including iodine deficiency in some areas in Southern Belarus, may have contributed to the promotion of thyroid tumorigenesis, but are not likely to be primarily responsible for this phenomenon. Because of the extremely low incidence of sporadic thyroid carcinoma in children before the accident, as well as in those born after 1986, it can be assumed that almost all cases of thyroid carcinomas in children from the affected areas are radiation-associated. Thus, this tragic disaster has created a unique human model of radiation-induced tumorigenesis in the thyroid gland, which becomes a significant tool for further studies of the mechanisms of radiation carcinogenesis.

Epidemiology, Clinical, and Histopathological Characteristics

Histologically, more than 98% of post-Chernobyl thyroid cancers are papillary carcinomas. The characteristic microscopical feature of these tumors is a high prevalence of solid pattern growth in tumor nodule, which appears as solid nests of malignant epithelial cells surrounded by varying amounts of fibrotic stroma. These contrast with the typical papillary growth characteristic of sporadic tumors. Thus, among 84 unselected cases operated on in Belarus between 1991 and 1992, the solid variant tumors (containing exclusively or predominantly solid areas of growth) were the most frequently observed, composing 34% of all tumors. Another 12% of tumors contained solid areas mixed with follicular and/or typical papillary areas, making the occurrence of solid growth pattern in this group even higher. On the other hand, the solid variant of papillary carcinoma is a rare finding in the spontaneous adult and pediatric population. Comparative analysis of pediatric papillary carcinomas from Belarus and two different regions in the United States revealed that the solid variant was present in 37% of radiation-induced, but only in 4% of sporadic tumors (Nikiforov et al., 1997). Similar data were obtained when post-Chernobyl tumors from the Ukraine were compared with those in the United Kingdom (Bogdanova et al., 1995).

Post-Chernobyl thyroid tumors also differ from spontaneous ones in their sex distribution. The female:male ratio was significantly lower among radiation-induced tumors, as it was 1.6:1 among 427 children and adolescents from Belarus, in contrast with 2.5:1 among 346 unexposed patients of the same age from Italy and France (Pacini et al., 1996).

Radiation-induced thyroid cancer, like other radiation-associated solid neoplasms, is known to require a certain period of time between radiation exposure and the detection of the first cases, referred to as the minimum induction period. Thus, among Japanese atomic bomb survivors, and exposed residents of the Marshall Islands after the testing of a thermonuclear device, thyroid cancer appeared 10 years or more after exposure (Prentice et al., 1982; Conard, 1984). A number of studies of pediatric thyroid cancer following therapeutic irradiation to the head and neck demonstrated an average time between X-ray therapy and tumor development varying from 6.3 to 9.6 years (reviewed in Nikiforov and Gnepp, 1994). At the same time, among Belarussian children, the increased rate of thyroid cancer was recognized as early as three years after the accident, and was quite obvious after four years. This short latent period is an unusual feature of post-Chernobyl tumors.

The age of the children at the time of exposure was an important modulating factor. Thus, for the series of tumors operated on in 1991–1992, one patient may have been exposed *in utero*, whereas the age of the others ranged from one month to 12 years (Nikiforov et al., 1996a). The highest number of patients that subsequently developed thyroid carcinomas was less than one year of age at the time of the Chernobyl accident, and this number decreased progressively through age 12, showing a strong inverse correlation between these two parameters (Figure 2). Conversely, none of the patients with benign thyroid lesions, which also increased in children from the affected areas, was less than two years of age at the time of exposure, and the number of cases increased as a function of age at exposure. An exposure age of five to six years was the threshold separating a statistically significant prevalence of malignant tumors in younger children from the more frequent benign lesions in older patients (Figure 2). One of the possible explanations for the critical importance of the age at exposure may be related to dose differences, since 52% of children with carcinomas and only 24% with benign lesions in this series were residents of Gomel, the most contaminated region (Nikiforov et al., 1996a).

The tissue environment in which thyroid carcinomas arise can provide information on the possible nature of premalignant lesions that may

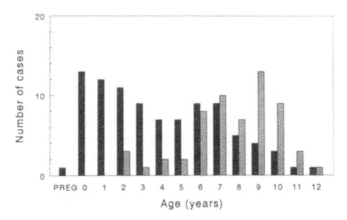

Figure 2. Age distribution of children who subsequently developed malignant and benign thyroid lesions at the time of the Chernobyl accident. Black bars, patients that subsequently developed thyroid carcinomas; hatched bars, patients that subsequently developed benign lesions as primary diagnosis. From Nikiforov et al. (1995) with permission.

serve as precursors for radiation-induced cancers. Thus, focal micro-papillary hyperplasia with graded degrees of severity was found frequently in these thyroid glands (Nikiforov et al., 1996a). Early stage lesions—delicate micropapillae—were found with a high prevalence in thyroid tissue surrounding malignant and benign nodules. Advanced papillae were present in far fewer cases. Even more rarely, branched papillae were observed, and in these few cases only one lesion was identified per gland; epithelial cells covering these papillae exhibited some cytological features consistent with a neoplastic epithelium. This fits with a concept of clonal progression, and it was hypothesized that these micropapillae are precursors for papillary thyroid carcinoma in post-Chernobyl radiation-associated tumors, with the following sequence: delicate papillae → advanced papillae → branched papillae → papillary carcinoma (Nikiforov et al., 1996a). However, further studies are needed to test this hypothesis of step-wise progression to papillary carcinoma.

Molecular Genetics

In the first sections of this review, we discussed the present understanding of the types of DNA damage associated with radiation. Although large-scale genetic rearrangements are a well-recognized sequelae of radiation exposure, more discrete abnormalities, such as point mutations and small deletions, are also observed. There is considerable information on the prevalence of mutations of specific oncogenes and tumor suppressor genes in sporadic thyroid cancers, some of which are known to be disrupted by rearrangements (i.e., *ret*/PTC), and others by simple base substitutions (i.e., *ras*, p53). Some of the initial studies on the Chernobyl tumors have focused on the genes known to play a role in the sporadic (i.e., non-radiation induced) forms of the disease.

Ras

Point mutations of the three *ras* genes have been reported in 18–62% of papillary thyroid carcinoma, and with even greater prevalence in follicular thyroid tumors (reviewed in Fagin, 1996). Even more importantly, several investigators have proposed a role for *ras* mutations in radiation-induced thyroid tumors. Lemoine et al. (1988) reported that DNA from 60% of radiation-induced rat thyroid tumors transfected into NIH3T3 cells scored positive in the nude mouse tumorigenesis assay, and in eight of nine cases transforming activity was due to activation of K-*ras*. In thyroid tumors from 12 patients exposed to therapeutic external radiation, 4 *ras* mutations were found, 3 of which were also in K-*ras* (Wright et al., 1991). However, in should be noted that most of these cases were follicular neoplasms rather than papillary carcinomas. In a recent report, 35% of 17 thyroid carcinomas, and 25% of 16 thyroid adenomas from patients with a history of external irradiation had *ras* mutations; four of six mutations in papillary carcinomas were in the K-*ras* gene (Challeton et al., 1995).

The prevalence of *ras* mutation in post-Chernobyl tumors was studied in a series of 33 papillary carcinomas, one follicular carcinoma, and 22 benign nodular lesions from Belarus (Nikiforov et al., 1996b). In contrast with previous reports, post-Chernobyl pediatric papillary carcinomas did not exhibit mutations of either N-, H-, or K-*ras*, indicating that in this series *ras* oncogenes are not a mediator of radiation-induced tumor formation (Figure 3). However, *ras* point mutations were detected in 1/1 case of follicular carcinoma (N-*ras* codon 61 $CAA^{gln} \rightarrow AAA^{lys}$), and in 3/7 follicular adenomas (N-*ras* codon 61 $CAA^{gln} \rightarrow CGA^{arg}$-two, $CAA^{gln} \rightarrow AAA^{lys}$), suggesting a possible role of this genetic event in the origin of follicular thyroid tumors. Indeed, it is conceivable that radiation may induce at least two distinct pathways of tumor initiation, and that *ras*

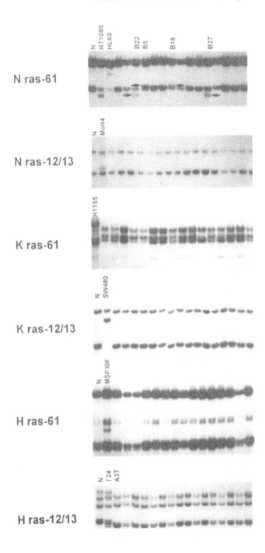

Figure 3. SSCP analysis for *ras* gene mutations in radiation-induced thyroid tumors. N: human placental DNA. HT1080, HL60, Molt4, H1155, SW480, MSF10F, T24, A3T: positive control human tumor cell lines. Unmarked lanes correspond to thyroid tumor samples. B22, B5, B18, and B27 are human thyroid tumor DNA samples containing strand shifts, which indicate the presence of mutations of N-*ras* codon 61 in these tumors. These were subsequently confirmed by sequencing. B22, follicular carcinoma; B5, B18, and B27, follicular adenomas. From Nikiforov et al. (1996b) with permission.

activation through point mutations may predispose to formation of follicular neoplasms.

Ret

Recent studies indicate that rearrangements resulting in the aberrant expression of the intracellular domain of the *ret* receptor (i.e., *ret*/PTC rearrangements) are very common in post-Chernobyl papillary thyroid carcinomas (Fugazzola et al., 1995; Klugbauer et al., 1995, Nikiforov et al., 1997). These three series consistently show that prevalence of *ret*/PTC rearrangements is high in post-Chernobyl pediatric papillary carcinomas, being found in 67–77% of these tumors. *Ret* rearrangements are formed by the fusion of the tyrosine kinase domain of c-*ret* with different 5' gene fragments resulting in inappropriate constitutive expression of the truncated intracellular domain of the receptor. It was proposed that this genetic event might reflect a radiation origin of these tumors. Indeed, the incidence found in pediatric post-Chernobyl tumors is much higher than that previously reported in the general (mostly adult) population, where three types of *ret* rearrangements, *ret*/PTC1, *ret*/PTC2, and *ret*/PTC3 have been detected in up to 34% of papillary thyroid carcinomas (summarized in Sugg et al., 1996). However, a recent study demonstrates a similarly high incidence of *ret*/PTC rearrangements in spontaneous thyroid carcinomas in Italian children, but not in adults, suggesting that the high prevalence of *ret* rearrangements might be an age-specific, rather than radiation-specific event (Bongarzone et al., 1996). On the other hand, the frequency of specific types of rearrangement appeared to be significantly different between radiation-induced and spontaneous tumors. Thus, a comparative analysis showed that among post-Chernobyl tumors *ret*/PTC3 rearrangement was found in 58%, *ret*/PTC1 in 16%, and *ret*/PTC2 in 3%, while among sporadic tumors *ret*/PTC1 was found in 47%, and *ret*/PTC3 in 18% (Nikiforov et al., 1997). These data suggest that the *ret*/PTC3 type of rearrangement is characteristic of pediatric papillary carcinomas from children exposed to radiation, in contrast to those occurring sporadically. Furthermore, solid variants have a high prevalence of *ret*/PTC3, whereas typical papillary carcinomas do not, suggesting that the different types of *ret* rearrangement confer neoplastic thyroid cells with distinct phenotypic properties. It is unclear whether *ret* rearrangements are a direct result of radiation DNA damage, or if they occur later in the evolution of the neoplastic clone. Ito et al. (1993) demonstrated occurrence of *ret*/PTC1 rearrangements in a culture of human undifferentiated thyroid carcinoma cells and fibrosarcoma cells harvested 48 hours after X-irradiation in a dose of 50 and 100 Gy. However, these data were obtained on clonal carcinoma cells growing *in vitro*, using very high levels of radiation, and it is premature to extrapolate these findings to the *in vivo* setting.

p53

Inactivating point mutations of the p53 tumor suppressor gene are highly prevalent in anaplastic and poorly-differentiated thyroid tumors, but not in well-differentiated papillary or follicular carcinomas (reviewed in Fagin, 1996). However, as discussed above, thyroid papillary carcinomas in children after Chernobyl are characterized by an unexpectedly high prevalence of solid growth tumors, considered by some authors as evidence of a more malignant phenotype. In addition, p53 mutations have been reported with a high frequency in radon-associated lung cancers from uranium miners (Vahakangas et al., 1992; Taylor et al., 1993). On the other hand, as discussed above, a rapid increase in wild-type p53 expression has been linked to the G1 arrest induced by ionizing radiation.

In a series of 33 post-Chernobyl papillary carcinomas, somatic point mutations of p53 were found in two tumors, with one missense mutation (exon 5, codon 160 ATG^{met}→GTG^{val}), and another silent mutation (codon 182, TGC^{cys}→TGT^{cys}) (Nikiforov et al., 1996b). Immunohistochemically, focally positive p53 staining was found in four papillary carcinomas being primarily confined to solid and poorly-differentiated areas in tumors. More recently, absence of p53 mutations was also reported in another series of 34 tumors from Belarus and Ukraine (Williams, 1996). These data indicate that inactivation of the p53 tumor suppressor gene is not involved in radiation-induced papillary thyroid tumor formation. As p53 inactivation is a late event in the progression of sporadic thyroid tumors, it is possible that structural abnormalities of p53 may occur in more advanced forms of the disease that may only become apparent in the future.

Genetic Instability

As discussed at length in the second section, several investigators have proposed that DNA damage after radiation may be secondary to induction of genomic instability, that in turn would lead to the gradual accumulation of genetic mutations. One or more of these would eventually lead to initiation or expansion of a neoplastic clone. Microsatellite instability is now recognized as a precursor of certain solid neoplasms (not radiation-induced), and is due to loss of function of one of the DNA mismatch repair enzymes. It has been proposed that putative delayed mechanisms of radiation-induced carcinogenesis may be associated with microsatellite instability. This hypothesis was tested on 15 radiation-induced post-Chernobyl papillary carcinomas from children utilizing 26 microsatellite repeat markers (Nikiforov et al., 1996c). Microsatellite mutations were found in only one case in a single locus of trinucleotide repeats. Furthermore, none of these tumors had loss of heterozygosity of any of the four DNA mismatch repair genes (hMSH2, hMLH1, hPMS1, and PMS2), or of the AT gene, indicating that both alleles of all these DNA repair enzymes were present in all tumors. These data suggest that the delayed effects of radiation in thyroid cell transformation *in vivo* are not likely to be due to microsatellite instability and mismatch DNA repair deficit, and that the loci of the major enzymes involved in this repair process are not subject to allelic losses.

The absence of microsatellite instability in these tumors, however, does not exclude a role for other forms of genome destabilization in the pathogenesis of radiation-induced thyroid tumors. Thus, a recent report demonstrates a higher frequency of germline mutations at human minisatellite loci in children from parents living in contaminated areas in Belarus (Dubrova et al., 1996). The mechanisms accounting for the radiation induced mutations at minisatellite loci are not known, but are probably distinct from alterations in DNA mismatch repair. At the present time, we are screening a series of post-Chernobyl papillary carcinomas for possible presence of minisatellite instability as a somatic event that may predispose to tumor formation or progression.

CONCLUSION

The tragic accident at Chernobyl has resulted in the most clearly defined epidemiological connection between radiation exposure and solid tumor formation in humans. As such, it provides a unique opportunity to explore the mechanisms of radiation-induced tumorigenesis. Based on present information on the biological effects of radiation, the ultimate nature of the genetic events giving rise to the papillary carcinomas in the ex-

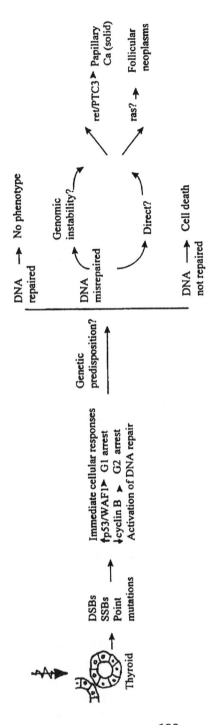

Figure 4. Hypothetical scheme of possible mechanisms of thyroid carcinoma formation after radiation exposure. Radiation causes DNA damage, and initiates a cascade of events leading to cell cycle arrest and activation of DNA repair. Failure of DNA repair can result in the generation of oncogenic sequences. This may occur either as a direct effect of radiation or indirectly through radiation-induced genomic instability, and a heightened tendency to accumulate gene mutations. *Ret/PTC3* rearrangements are the hallmark of radiation-induced papillary carcinomas. Point mutations of *ras* are found in many radiation-associated thyroid adenomas and may also be radiation-induced (although rigorous comparisons between radiation-related and spontaneously arising follicular neoplasms in age-matched populations have not been

posed children may have resulted either directly from DNA damage to oncogenic sequences (i.e., *ret*/PTC), or may be due to downstream events taking place after the genome of the affected cell has been destabilized by as yet unexplained mechanisms (Figure 4). It is now clear that radiation-induced papillary carcinomas of Chernobyl have certain "signature" genetic features, namely a high prevalence of a particular form of *ret*/PTC rearrangement. This provides a definable end-point for future studies of radiation-induced DNA damage in thyroid cells, and may allow for the testing of more targeted hypothesis as to the natural history of these tumors.

ACKNOWLEDGMENTS

This work was supported in part by grant CA 50706 and CA 72597, and a University of Cincinnati Cancer Research Challenge award. J.A.F. is the recipient of an Established Investigatorship Award from the American Heart Association and Bristol Myers-Squibb.

REFERENCES

Akslen, L.A., Haldorsen, T., Thoresen, S.O., & Glattre, E. (1990). Incidence of thyroid cancer in Norway 1970-1985. Population review of time trend, sex, age, histological type and tumor stage in 2625 cases. APMIS 98, 549-558.

Astakhova, L.N., Demidchuk, E.P., Davydova, E.V., Arinchin, A.N., Gres', N.A., Zelenko, S.M. et al. (1991). Health status of Byelorussian children and adolescents exposed to radiation as consequence of the Chernobyl AES accident (in Russian). Vestnik Akademii Meditsinskikh Nauk Sssr 11, 25-27.

Baverstock, K., Egloff, B., Pinchera, A., Ruchti, C., & Williams, D. (1992). Thyroid cancer after Chernobyl. Nature (London) 359, 21-22.

Becker, D.V., Robbins, J., Beebe, G.W., Bouville, A.C., & Wachholz, B.W. (1996). Childhood thyroid cancer following the Chernobyl accident. A status report. Endocrinol. Metab. Clin. North Amer. 25, 197-211.

Bernhard, E.J., Maity, A., Muschel, R.J., & McKenna, W.G. (1995). Effects of ionizing radiation on cell cycle progression. Radiat. Environ. Biophys. 34, 79-83.

Blakely, W.F., Joner, E.I., & Ward, J.F. (1982) Repair of DNA damage in Chinese hamster V79 cells after varying Radiation doses. Radiat. Res. (Abstract). 91, 387.

Bogdanova, T., Bragarnik, M., Tronko, N.D., Harach, H.R., Thomas, G.A., & Williams, E.D. (1995). Thyroid cancer in the Ukraine post Chernobyl. Thyroid 5, (Suppl. 1), S-28.

Bongarzone, I., Fugazzola, L., Vigneri, P., Mariani, L., Mondellini, P., Pacini, F., Basolo, F., Pinchera, A., Pilotti, S., & Pierotti, M.A. (1996). Age-related activation of the tyrosine kinase receptor protooncogenes RET and NTRK1 in papillart thyroid carcinoma. J. Clin. Endocrinol. Metab. 81, 2006-2009.

Challeton, C., Bounacer, A., Du Villard, J.A., Caillou, B., De Vathaire, F., Monier, R., Schlumberger, M., & Suarez, H.G. (1995). Pattern of ras and gsp oncogene mutations in radiation-induced human thyroid tumors. Oncogene 11, 601-603.

Chen, D.J., Park, M.S., Campbell, E., Oshimura, M., Liu, P., Zhao, Y., White, B.F., & Siciliano, M.J. (1992). Assignment of a human DNA double-strand break repair gene (XRCC5) to chromosome 2. Genomics 13, 1088-1094.

Collins, A.R. (1993). Mutant rodent cell lines sensitive to ultraviolet light, ionozing radiation and cross-linking agents: a comprehensive survey of genetic and biochemical characteristics. Mutat. Res., DNA Repair 293, 99-118.

Conard, R.A. (1984). In: Radiation Carcinogenesis: Epidemiology & Biological Significance. (Boice, J.D., & Fraumeni, J.F., eds.), pp. 57-70. Raven Press, New York.

Cox, R., Masson, W.K., Weichselbaum, R.R., Nove, J., & Little, J.(1981). Repair of potentially lethal damage in X-irradiated cultures of normal and ataxia-telangiectasia human fibroblasts. Int. J. Radiat. Biol. 39, 357-365.

Cronkite, E.P., Bond, V.P., & Conard, R.A. (1995). Medical effects of exposure of human beings to fallout radiation from a thermonuclear explosion. Stem Cells 13, 49-57.

De Keyser, L.F.M., & Van Herle, A.J. (1985). Differentiated thyroid cancer in children. Head & Neck Surg. 8, 100-114.

Dubrova, Y.E., Jeffreys, A.J., & Malashenko, A.M. (1993). Mouse minisatellite mutations induced by ionizing radiation. Nature Genet. 5, 92-94.

Dubrova, Y.E., Nesterov, V.N., Krouchinsky, N.G., Ostapenko, V.A., Neumann, R., Neil, D.L., & Jeffreys, A.J. (1996). Human minisatellite mutation rate after the Chernobyl accident. Nature (London) 380, 683-686.

Duffy, B.J., & Fitzgerald, P.J. (1950). Cancer of the thyroid in children: a report of 28 cases. J. Clin. Endocrinol. Metab. 10, 1296-308.

Ezaki, H., Takeichi, N., Yoshimoto, Y. (1991). Thyroid cancer: epidemiological study of thyroid cancer in A-bomb survivors from extended life span study cohort in Hiroshima. J. Radiat. Res. 32 (Suppl.) 193-200.

Fagin, J.A. (1996). In: Werner and Ingbar's. The Thyroid. A Fundamental and Clinical Text. (Braverman, L.E., & Utiger, R.D., eds.), pp. 909-915, Lippincott-Raven, Philadelphia-New York.

Felber, M., Burns, F.J., & Garte, S.G. (1994). DNA fingerprinting analysis of radiation-induced rat skin tumors. Cancer Biochem. Biophys. 14, 163-170.

Fornace, Jr., A.J. (1992). Mammalian genes induced by radiation; activation of gene associated with growth control. Annu. Rev. Genet. 26, 507-526.

Fugazzola, L., Pilotti, S., Pinchera, A., Vorontsova, T.V., Mondellini, P., Bongarzone, I., Greco, A., Astakhova, L., Butti, M.G., Demidchik, E.P., Pacini, F., & Pierotti, M.A. (1995). Oncogenic rearrangements of the RET proto-oncogene in papillary thyroid carcinomas from children exposed to the Chernobyl nuclear accident. Cancer Res. 55, 5617-5620.

Giaccia, A.J., Denko, N., MacLaren, D., Mirman, D., Waldren, C., Hart, I., & Stamato, T.D. (1990). Human chromosome 5 complements the DNA double-strand break-repair deficiency and gamma-ray sensitivity of the XR-1 hamster variant. Am. J. Hum. Genet. 47, 459-469.

Giver, C.R., Nelson Jr, S.L., Cha, M.Y., Pongsaensook, P., & Grosovsky, A.J. (1995). Mutational spectrum of X-ray induced TK⁻ human cell mutants. Carcinogenesis 16, 267-275.

Goodhead, D. T. (1989). The initial physical damage produced by ionizing radiation. Int. J. Radiat. Biol. 56, 623-634.

Grosovsky, A.J., Drobetsky, E.A., deJong, P.J., & Glickman, B.W. (1986). Southern analysis of genomic alterations in gamma-ray-induced APRT⁻ hamster cell mutants. Genetics 113, 405-415.

Grosovsky, A.J., de Boer, J.G., deJong, P.J., Drobetsky, E.A., & Glickman, B.W. (1988). Base substitutions, frameshifts, and small deletions constitute ionizing radiation-induced point mutations in mammalian cells. Proc. Natl. Acad. Sci. USA 85, 185-188.

Ilyin, L.A., Balonov, M.I., Buldakov, L.A., Bur'yak, V.N., Gordeev, K.I., Dement'ev, S.I., Zhakov, I.G., Zubovsky, G.A., Kondrusev, A.I., Konstantinov, Y.O., Linge, I.I., Likhtarev, I.A., Lyaginskaya, A.M., Matyuhin, V.A., Pavlovsky, O.A., Potapov, A.I., Prysyazhnyuk, A.E., Ramsaev, P.V., Romanenko, A.E., Savkin, M.N., Starkova, N.T., Tron'ko, N.D., & Tsyb, A.F. (1990). Radiocontamination patterns and possible health consequences of the accident at the Chernobyl nuclear power station. J. Radiol. Prot. 10, 13-29.

International Conference "One decade after Chernobyl". European Commission, International Atomic Energy Agency, World Health Organization. Vienna, Austria, 1996.

Ito, T., Seyama, T., Iwamoto, K.S., Hayashi, T., Mizuno, T., Tsuyama, N., Dohi, Nakamura, N., & Akiyama, M. (1993). *In vitro* irradiation is able to cause *RET* oncogene rearrangement. Cancer Res. 53, 2940-2943.

Kadhim, M.A., Lorimore, S.A., Townsend, K.M.S., Goodhead, D.T., Buckle, V.J., & Wright, E.G. (1995). Radiation-induced genomic instability: delayed cytogenetic aberrations and apoptosis in primary human bone marrow cells. Int. J. Radiat. Biol. 67, 287-293.

Karran, P. (1995). Appropriate partners make good matches. Science. 268, 1857-1858.

Kastan, M.B., Onyekwere, O., Sidransky, D., Vogelstein, B.,& Craig, R.W. (1991). Participation of p53 protein in the cellular response to DNA damage. Cancer Res., 51, 6304-6311.

Kastan, M.B., Zhan, Q., El-Deiry, W.S., Carrier, F.,Jacks, T., Walsh, W.V., Vogelstein, B., Plunkett, B.S., & Fornace, Jr. A.J. (1992). A mammalian cell cycle checkpoint pathway utilising p53 and GADD45 is defective in ataxia-telangiectasia. Cell 71, 587-597.

Kazakov, V.S., Demidchik, E.P., & Astakhova, L.N. (1992). Thyroid cancer after Chernobyl. Nature (London) 359, 21.

Klugbauer, S., Lengfelder, E., Demidchik, E.P., & Rabes, H.M. (1995). High prevalence of RET rearrangement in thyroid tumors of children from Belarus after the Chernobyl reactor accident. Oncogene 11, 2459-2467.

Kovacs, M.S., Evans, J.W., Johnstone, I.M., & Brown, J.M. (1994). Radiation-induced damage, repair and exchange formation in different chromosomes of human fibroblasts determined by fluorescence *in situ* hybridization. Radiat. Res. 137, 34-43.

Kronenberg, A. (1994). Radiation-induced genomic instability. Int. J. Radiat. Biol. 66, 603-609.

Kuebritz, S.J., Plunkett, B.S., Walsh, W.V., & Kastan, M.B. (1992). Wild-type p53 is a cell cycle checkpoint determinant followinr radiation. Proc. Natl. Acad. Sci. USA 89, 7491-7495.

Lemoine, N.R., Mayall, E.S., Williams, E.D., Thurston, V., & Wynford-Thomas, D. (1988). Agent-specific ras oncogene activation in rat thyroid tumours. Oncogene 3, 541-544.

Little, J.B. (1994). Failla Memorial Lecture:Changing Views of Cellular Radiosensitivity. Radiat. Res. 140, 299-311.

Medical consequences of the catastrophe at the Chernobyl AES in Belarus. (1996). In: Ecological, Medical-Biological and Social-Economical Consequences of the Catastrophe at the Chernobyl AES in Belarus (Konoplia, E.F., & Rolevitch, I.B., eds.), pp. 103-214. Ministry for Emergency Situations and Protection of Population from the Consequences of the Catastrophe at the Chernobyl AES of the Republic of Belarus & Institute of Radiobiology of Academy od Science of Belarus, Minsk.

Muris, D.F.R., Bezzubova, O., Buerstedde, J.-M., Vreeken, K., Balajee, A.S., Osgood, C.J., Troelstra, C., Hoeijmakers, J.H.J., Ostermann, K., Schmidt, H., NataraJan A.T., Eeken, J.C.J., Lohman, P.H.M., & Pastink, A. (1994). Cloning of human and mouse genes homologous to RAD52, a yeast gene involved in DNA repair and recombination. Mutat. Res., DNA Repair 315, 295-305.

Murnane, J.P., & Kapp, L.N. (1993). A critical look at the association of human genetic syndroms with sensitivity to ionizing radiation. Semin. Cancer Biol. 4, 93-104.

Nagasawa, H., Li, C.-Y., Maki, C.G., Imrich, A.C., & Little, J.B. (1995). Relationship between radiation-induced G_1 phase arrest and p53 function in human tumor calls. Cancer Res. 55, 1842-1846.

Namba, H., Hara, T., Tukazaki, T., Migita, K., Ishikawa, N., Ito, K., Nagataki, S., & Yamashita, S. (1995). Radiation-induced G_1 arrest is selectively mediated by the p53-WAF/Cip1 pathway in human thyroid cells. Cancer Res. 55, 2075-2080.

Nelson, S.L., Giver, C.R., & Grosovsky, A.J. (1994). Spectrum of X-ray induced mutations in the human hprt gene. Carcinogenesis 15, 495-502.

Nikiforov, Y., & Gnepp, D.R. (1994). Pediatric thyroid cancer after the Chernobyl disaster. Pathomorphologic study of 84 cases (1991-1992) from the Republic of Belarus. Cancer 74, 748-766.

Nikiforov, Y., Heffess, C.S., Korzenko, A.V., Fagin, J.A., & Gnepp, D.R. (1995). Characteristics of follicular tumors and non-neoplastic thyroid lesions in children and adolescents exposed to radiation after the Chernobyl disaster. Cancer 76, 900-909.

Nikiforov, Y., Gnepp, D.R., & Fagin, J.A. (1996a). Thyroid lesions in children and adolescents after the Chernboyl disaster:Implications for the study of radiation carcinogenesis. J. Clin. Endocrinol. Metab. 81, 9-14.

Nikiforov, Y.E., Nikiforova, M., Gnepp, D.R., & Fagin, J.A. (1996b). Prevalence of mutations of ras and p53 in thyroid tumors from children exposed to radiation after Chernobyl. Oncogene 13, 687-693.

Nikiforov, Y.E., Brenta, G., Fagin, J.A. (1996c). Genomic Instability in Radiation-Induced and Sporadic Thyroid Tumors: Thyroid Cancers Exhibit

Large-Scale Genomic Abnormalities but Not Microsatellite Instability, **p.** 3-824. The Endocrine Society, 78th Annual Meeting (Abstract).

Nikiforov, Y.E. & Fagin, J.A. (1997). Risk factors for thyroid cancer. Trends in Endocrinol. Metab. 8. In press.

Nikiforov, Y.E., Rowland, J.M., Bove, K., Monforte-Munoz, H., & Fagin, J.A. (1997). Distinct pattern of *ret* oncogene rearrangements in morphological variants of radiation-induced and sporadic thyroid papillary carcinomas in children. Cancer Res. 57, 1690-1694.

North, P., Ganesh, A., & Thacer, J. (1990). The rejoining of double-strand breaks in DNA by human cell extracts. Nucleic Acids Res. 18, 6205-6210.

Okeanov, A.E., & Averkin, Y.I. (1991). In: Catastrophe at the Chernobyl AES and Estimation of Health State of Population of the Republic of Belarus (Matyukhin, V.A., Astakhova, L.N., Konigsberg, Y.E., & Nalivko, S.N., eds.), pp. 25-33. Research Institute of Radiation Medicine, Minsk.

Pacini, F., Vorontsova, T., Demidchik, E., Rogna, G., Schlumberger, M., Agate, L., Molinaro, E., Mancusi, F., Romei, C., Filesi, M., Cherstvoy, E., & Pinchera, A. (1996). Clinical-epidemiological features of post-Chernobyl thyroid carcinomas in Belarus children and adolescents as compared to control cases from Italy and France. (Abstract) J. Endocrinol. Invest. 19, (Suppl 6), 60.

Paquette, B., & Little, J.B. (1994). *In vivo* enhancement of genomic instability in minisatellite sequences of mouse C3H/10T1/2 cells transformed *in vitro* by X-rays. Cancer Res. 54, 3173-3178.

Polyanskaya, O.N., Astakhova, L.N., Drozd, V.M., Markova, S.V., Dubovtsov, A.M., & Mityukova, T.A. (1991). In: Catastrophe at the Chernobyl AES and Estimation of Health State of Population of the Republic of Belarus (Matyukhin, V.A., Astakhova, L.N., Konigsberg, Y.E., & Nalivko, S.N., eds.), pp. 62-69. Research Institute of Radiation Medicine, Minsk.

Prentice, R.L., Kato, H., Yoshimoto, K., & Maso, M. (1982). Radiation exposure and thyroid cancer incidence among Hiroshima and Nagasaki residents. Monogr. Natl. Cancer Inst. 62, 207-212.

Sadamoto, S., Suzuki, S., Kamiya, K., Kominami, R., Dohl, K., & Niwa, O. (1994). Radiation induction of germline mutation at a hypervariable mouse minisatellite locus. Inter. J. Radiat. Biol. 65, 549-557.

Shibata, D., Peinado, M.A., Ionov, Y., Malkhosyan, S., & Perucho, M. (1994). Genomic instability in repeated sequences is an early somatic event in colorectal tumorigenesis that persists after transformation. Nature Genet. 6, 273-281.

Shinohara, A., Ogava, H., Matsuda, Y., Ushio, N., Ikeo, K., & Ogawa, T. (1993). Cloning of human, mouse and fission yeast recombination genes homologous to *RAD51* and *recA*. Nature Genet. 4, 239-243.

Stsjazhko, V.A., Tsyb, A.F., tronko, N.D., Souchkevitch, G., & Baverstock, K.F. (1995). Childhood thyroid cancer since accident at Chernobyl. Brit. Med. J. 310, 801.

Sugg, S.L., Zheng, L., Rosen, I.B., Freeman, J.L., Ezzat, S., & Asa, S.L. (1996). ret/PTC-1, -2, and -3 oncogene rearrangements in human thyroid carcinomas: implication for metastatic potential? J. Clin. Endocrinol. Metab. 81, 3360-3365.

Taylor, A.M.R., Harnden, D.G., Arlett, C.F., Harcourt, S.A., Lehmann, A.R., Stevens, S., & Bridges, B.A. (1975). Ataxia-telangiectasia: a human mutation with abnormal radiation sensitivity. Nature (London) 258, 427-429.

Taylor, J.A., Watson, M.A., Devereux, T.R., Michels, R.Y., Saccomanno, G., & Anderson, M. (1993). Mutational hotspot in the p53 gene in radon-associated lung tumors from uranium miners. Lancet 343, 86-87.

Tebbs, R.S., Zhao, Y., Tucker, J.D., Schreerer, J.B., Siciliano, M.J., Hwang, N., Liu, N., Legerski, R.J., & Thompson L.N. (1995). Correction of chromosomal instability and sensitivity to diverse mutagens by cDNA of the XRCC3 DNA repair gene. Proc. Nat. Acad. Sci. USA 92, 6354-6358.

Terzaghi, M., & Little, J.B. (1976). X-radiation induced transformation in a C3H mouse embryo-derived cell line. Cancer Res. 36, 1367-1374.

Thacker, J. (1986). The nature of mutants induced by ionising radiation in cultured hamster cells. Mutation Res. 160, 267-275.

Thompson, L.H., Brookman, K.W., Jones, N.J., Allen, S.A., & Carrano, A.V. (1990). Molecular cloning of the human XRCC1 gene, which correct defective DNA strand-break repair and sister chromatid-exchange. Mol. Cell. Biol. 10, 6160-6171.

Vahakangas, K.H., Samet, J.M., Metcalf, R.A., Welch, J.A., Bennett, W.P., Lane, D.P., & Harris, C.C. (1992). Mutations of p53 and ras genes in radon-associated lung cancer from uranium miners. Lancet 339, 576-580.

Ward, J.F. (1988). DNA damage produced by ionizing radiation in mammalian cells: Identities, mechanisms of formation and repairability. Progr. Nucleic Acids Mol. Biol. 35, 95-125.

Ward, J.F. (1994). DNA damage as the cause of ionizing radiation-induced gene activation. Radiat. Res. 138, S85-S88.

Weinert, T.A., & Hartwell, L.H. (1988). The RAD9 gene controls the cell cycle response to DNA damage in Saccharomyces cerevisiae. Science 241, 317-322.

Williams, E.D. (1996). Pathology and molecular biology of post Chernobyl thyroid carcinoma (Abstract). J. Endocrinol. Invest. 6, (Suppl.), 4.

Winship, T., & Rosvoll, R.V. (1970). Thyroid Carcinoma in Childhood: Final Report on a 20 Year Study. Clin. Proc. Children's Hosp. Washington,DC 26, 327-349.

Wright, P.A., Williams, E.D., Lemoine, N.R., & Wynford-Thomas, D. (1991). Radiation-associated and 'spontaneous' human thyroid carcinomas show a different pattern of ras oncogene mutation. Oncogene 6, 471-473.

Young, J.L., Ries, L.G., Silverberg, E., Horm, J.W., & Miller, & R.W. (1986). Cancer incidence, survival, and mortality for children younger than 15 years. Cancer 58, 598-602.

Zdzienicka, M.Z. (1995). Mammalian mutants defective in the response to ionizing radiation-induced DNA damage. Mutat. Res., DNA Repair 336, 203-213.

Zhan, Q., Bae, I., Kastan, M.B., & Fornace, A.J. (1994). The p53-dependent γ-ray responce of GADD45. Cancer Res. 54, 2755-2760.

INDEX

Advances in Molecular and Cellular Endocrinology

Edited by **Derek Leroith**, *Diabetes Branch, NIDDK, National Institutes of Health, Bethesda, Maryland*

Volume 1, 1997, 265pp. $112.50/£72.50
ISBN: 0-7623-0158-9

JAI PRESS INC.

55 Old Post Road No. 2 - P.O. Box 1678
Greenwich, Connecticut 06836-1678
Tel: (203) 661- 7602   Fax: (203) 661-0792